ARISTOTLE'S *GENERATION OF ANIMALS*

Generation of Animals is one of Aristotle's most mature, sophisticated, and carefully crafted scientific writings. His overall goal is to provide a comprehensive and systematic account of how animals reproduce, including a study of their reproductive organs, what we would call fertilization, embryogenesis, and organogenesis. In this book, international experts present thirteen original essays providing a philosophically and historically informed introduction to this important work. They shed light on the unity and structure of the *Generation of Animals*, the main theses that Aristotle defends in the work, and the method of inquiry he adopts. They also open up new avenues of exploration of this difficult and still largely unexplored work. The volume will be essential for scholars and students of ancient philosophy as well as of the history and philosophy of science.

ANDREA FALCON is Associate Professor in the Department of Philosophy at Concordia University, Montréal. He is the author of *Corpi e Movimenti. La fortuna del De caelo nel mondo antico* (2001), *Aristotle and the Science of Nature: Unity without Uniformity* (Cambridge 2005); *Aristotelianism in the First Century BCE: Xenarchus of Seleucia* (Cambridge 2012); and *Aristotelismo* (2017). He is the editor of the *Brill's Companion to the Reception of Aristotle* (2016).

DAVID LEFEBVRE is Professor of Philosophy at the Université Clermont-Auvergne (UCA) in Clermont-Ferrand. His publications include a new translation of Aristotle's *Generation of Animals* (2014), and he is co-editor of *Dunamis. Autour de la puissance chez Aristote* (2008) and *La Métaphysique de Théophraste. Principes et apories* (2015) and author of *DYNAMIS. Sens et Genèse de la notion aristotélicienne de puissance* (2017).

CAMBRIDGE CRITICAL GUIDES

Recent Titles in this Series

Kant's Critique of Practical Reason
Edited by ANDREWS REATH and JENS TIMMERMANN

Wittgenstein's Philosophical Investigations
Edited by ARIF AHMED

Kierkegaard's Concluding Unscientific Postscript
Edited by RICK ANTHONY FURTAK

Plato's Republic
Edited by MARK L. MCPHERRAN

Plato's Laws
Edited by CHRISTOPHER BOBONICH

Spinoza's Theological-Political Treatise
Edited by YITZHAK Y. MELAMED and MICHAEL A. ROSENTHAL

Aristotle's Nicomachean Ethics
Edited by JON MILLER

Kant's Metaphysics of Morals
Edited by LARA DENIS

Nietzsche's On the Genealogy of Morality
Edited by SIMON MAY

Kant's Observations and Remarks
Edited by RICHARD VELKLEY and SUSAN SHELL

Augustine's City of God
Edited by JAMES WETZEL

Descartes' Meditations
Edited by KAREN DETLEFSEN

Kant's Religion within the Boundaries of Mere Reason
Edited by GORDON MICHALSON

Kant's Lectures on Anthropology
Edited by ALIX COHEN

Kierkegaard's Fear and Trembling
Edited by DANIEL CONWAY

Kant's Lectures on Ethics
Edited by LARA DENIS and OLIVER SENSEN

Aristotle's Physics
Edited by MARISKA LEUNISSEN

Aristotle's Politics
Edited by THORNTON LOCKWOOD and THANASSIS SAMARAS

Aquinas's Disputed Questions on Evil
Edited by M. V. DOUGHERTY

Plato's Symposium
Edited by PIERRE DESTRÉE and ZINA GIANNOPOULOU

Spinoza's Ethics
Edited by YITZHAK Y. MELAMED

Kant's Critique of Pure Reason
Edited by JAMES R. O'SHEA

Aristotle's Generation of Animals
Edited by ANDREA FALCON and DAVID LEFEBVRE

ARISTOTLE'S *GENERATION OF ANIMALS*

A Critical Guide

Edited by

ANDREA FALCON
Concordia University

DAVID LEFEBVRE
University of Clermont-Auvergne

CAMBRIDGE
UNIVERSITY PRESS

CAMBRIDGE
UNIVERSITY PRESS

University Printing House, Cambridge CB2 8BS, United Kingdom

One Liberty Plaza, 20th Floor, New York, NY 10006, USA

477 Williamstown Road, Port Melbourne, VIC 3207, Australia

314–321, 3rd Floor, Plot 3, Splendor Forum, Jasola District Centre, New Delhi – 110025, India

79 Anson Road, #06–04/06, Singapore 079906

Cambridge University Press is part of the University of Cambridge.

It furthers the University's mission by disseminating knowledge in the pursuit of education, learning, and research at the highest international levels of excellence.

www.cambridge.org
Information on this title: www.cambridge.org/9781107132931
DOI: 10.1017/9781316459386

First published 2018

Printed in the United Kingdom by Clays, St Ives plc

A catalogue record for this publication is available from the British Library.

Library of Congress Cataloging-in-Publication Data
NAMES: Falcon, Andrea, editor. | Lefebvre, David, editor.
TITLE: Aristotle's generation of animals : a critical guide / edited by Andrea Falcon, Concordia University, Montréal, David Lefebvre, Université de Paris I.
DESCRIPTION: Cambridge, United Kingdom ; New York, NY : Cambridge University Press, 2017. | SERIES: Cambridge critical guides | "The majority of the essays collected in this volume originate from papers delivered at the international conference on Aristotle's Generation of Animals that took place in Paris, January 9-11, 2014." | Includes bibliographical references and index.
IDENTIFIERS: LCCN 2017027451 | ISBN 9781107132931 (Hardback : alk. paper)
SUBJECTS: LCSH: Aristotle. On the generation of animals. | Reproduction. | Animals (Philosophy)
CLASSIFICATION: LCC QP251 .A674 2017 | DDC 612.6–DC23
LC record available at https://lccn.loc.gov/2017027451

ISBN 978-1-107-13293-1 Hardback

Contents

Contributors

ROBERT BOLTON is Professor in the Department of Philosophy at Rutgers University. He is the author of numerous articles on Aristotle, with a concentration on his method of scientific inquiry. Some of his articles are collected in R. Bolton, *Science, dialectique et éthique chez Aristote: essais d'épistémologie aristotélicienne* (2010). He is the editor, together with James G. Lennox, of *Being, Nature, and Life in Aristotle* (Cambridge University Press 2010).

CRISTINA CERAMI is Senior Research Fellow at the CNRS, Paris (UMR 7219: SPHERE/CHSPAM). She works on Aristotle and the Aristotelian tradition in the Greek and Arabic worlds, with a special concentration on Alexander of Aphrodisias and Averroes. She is the author of *Génération et substance: Aristote et Averroès entre physique et métaphysique* (2015). She is the editor of *Nature et sagesse: physique et métaphysique dans la tradition aristotélicienne. Recueil de textes en hommage à Pierre Pellegrin* (2015).

SOPHIA CONNELL is lecturer in Philosophy at Birkbeck College, University of London. She has been a fellow, admissions tutor, and director of studies in Philosophy at Selwyn College, Cambridge, and a bye-fellow and director of studies in Philosophy at Newnham College, Cambridge. She has written articles on Aristotelian biology, and Galen and Aristotle on sex difference. She is the author of *Aristotle on Female Animals: A Study in the* Generation of Animals (Cambridge University Press 2016).

ANDREA FALCON is Associate Professor in the Department of Philosophy at Concordia University, Montreal. He works on Aristotle and the Aristotelian tradition in antiquity, with a focus on Aristotle's philosophy of nature. He is the author of *Corpi e movimenti: la fortuna del* De caelo *nel mondo antico* (2001); *Aristotle and the Science of Nature: Unity without Uniformity* (Cambridge University Press 2005); *Aristotelianism in the First Century BCE: Xenarchus of Seleucia* (Cambridge University Press 2012); and *Aristotelismo* (2017). He is the editor of the *Brill's Companion to the Reception of Aristotle in Antiquity* (2016).

JESSICA GELBER is Assistant Professor in the Department of Philosophy at the University of Pittsburgh. Her current research is centered on foundational issues in ancient Greek science, particularly in Aristotle's biology, and has written about such topics as teleology, essence, and efficient causation in Aristotle's natural science. She completed her doctoral dissertation on Aristotle's *Generation of Animals* at the University of California at Berkeley.

ALLAN GOTTHELF (December 30, 1942–August 30, 2013) was Emeritus Professor of philosophy at The College of New Jersey and Anthem Foundation Distinguished Scholar at Rutgers University. He is the editor of *Aristotle on Nature and Living Things: Philosophical and Historical Studies* (1985) and, together with James G. Lennox, *Philosophical Issues in Aristotle's Biology* (Cambridge University Press 1987). His most important articles are collected in *Teleology, First Principles, and Scientific Method in Aristotle's Biology* (2012).

JOCELYN GROISARD, a former pupil of the École Normale Supérieure (Paris), has worked at the National Library of France in the section of Greek manuscripts before leaving for Japan, where he teaches French at the Tokyo Metropolitan University. He is the author of a translation of Aristotle's *Meteorology* (2008), a critical edition of Alexander of Aphrodisias' *On Mixture* [*De mixtione*] (*Sur la mixtion et la croissance*, 2013), and a monograph on ancient theories of mixture from Aristotle to Late Antiquity (*Mixis: le problème du mélange dans la philosophie grecque d'Aristote à Simplicius*, 2016).

DEVIN HENRY is Associate Professor of Philosophy at the University of Western Ontario. He received his PhD from King's College London, where he wrote a dissertation on the metaphysical foundations of Aristotle's account of biological generation. He is the author of several articles on topics in Aristotle's philosophy of science (including classification, teleology, inheritance, and sexism), as well as Plato's late epistemology. He is currently writing a book that examines the ways in which Aristotle's general theory of substantial generation is reflected in, and modified by, his more specific account of the generation of biological substance in Aristotle's *Generation of Animals*.

DAVID LEFEBVRE is Professor of Philosophy at the University Clermont-Auvergne (UCA) in Clermont-Ferrand. He has published on Aristotle and the Peripatos – Theophrastus, Strato of Lampsacus. He is the author of *Dynamis: sens et genèse de la notion aristotélicienne de puissance* (2017). His most recent publications include also a new translation of Aristotle's *Generation of Animals* (2014). He has co-edited *Dunamis: autour de la puissance chez Aristote* (2008), and *La Métaphysique de Théophraste: principes et apories* (2015).

JAMES G. LENNOX is Professor of History and Philosophy of Science at the University of Pittsburgh. He has published widely on Aristotle and Darwin. Some of his articles on Aristotle's biology are collected in J. G. Lennox, *Aristotle's Philosophy of Biology: Essays in the Origins of Life Science* (Cambridge University Press 2001). He has edited, together with Allan Gotthelf, *Philosophical Issues in Aristotle's Biology* (Cambridge University Press 1987); together with Marie-Louise Gill, *Self-Motion from Aristotle to Newton* (1995); together with Robert Bolton, *Being, Nature, and Life in Aristotle* (Cambridge University Press 2010). He has translated Aristotle's *Parts of Animals* for the Clarendon Aristotle Series (2001).

MARISKA LEUNISSEN is Associate Professor in the Department of Philosophy at the University of North Carolina, Chapel Hill. She works on Aristotle with a concentration on his philosophy of nature and its connections to his philosophy of science and his ethics and politics. She is the author of *Explanation and Teleology in Aristotle's Science of Nature* (Cambridge University Press 2010). She is the editor of *Aristotle's Physics: A Critical Guide* (Cambridge University Press 2015) and co-editor, with Frans J. de Haas and Marije Martijn, of *Interpreting Aristotle's* Posterior Analytics *in Late Antiquity and Beyond* (2010).

PIERRE PELLEGRIN is Director of Research Emeritus at the CNRS, Paris (UMR 7219: SPHERE/CHSPAM). He is the author of *La classification des animaux chez Aristote: statut de la biologie et unité de l'aristotélisme* (1982; English translation by Anthony Preus: *Aristotle's Classification of Animals: Biology and the Conceptual Unity of the Aristotelian Corpus*, 1986) and, together with Michel Crubellier, *Aristote: le philosophe et les savoirs* (2002). He is the editor, together with Daniel Devereux, of *Biologie, logique et métaphysique chez Aristote* (1990) and, together with Monique Canto-Sperber, of *Le style de la pensée: recueil de textes en hommage à Jacques Brunschwig* (2002). He has translated several works of Aristotle into French, including the *Physics* (2000), *Posterior Analytics* (2005), *Parts of Animals* (2011), and *History of Animals* (2014).

MARWAN RASHED is Professor of Greek and Arabic Philosophy at Paris-Sorbonne University. Among his most recent books are *De la génération et la corruption: édition, traduction et commentaire* (2005); *Essentialisme: Alexandre d'Aphrodise entre logique, physique et cosmologie* (2007); *Alexandre d'Aphrodise: commentaire perdu à la* Physique *d'Aristote (Livres IV–VIII)* (2011); *L'héritage aristotélicien: textes inédits de l'antiquité* (2014, second augmented edition).

GREGORY SALMIERI teaches at Rutgers University and is a philosophy fellow at the Anthem Foundation. He previously held research and teaching positons in philosophy at the University of North Carolina, Chapel Hill, and Boston University. He is editor, together with Allan Gotthelf, of *A Companion to Ayn Rand* (2016) and *Knowing and Coming to Know: Essays on Aristotle's Epistemology* (forthcoming).

Preface

The majority of the essays collected in this volume originate from papers delivered at the international conference on Aristotle's *Generation of Animals* that took place in Paris, January 9–11, 2014. The conference was jointly organized in Paris at the Sorbonne University by the late Allan Gotthelf and David Lefebvre under the title "Form and Necessity: On Aristotle's *Generation of Animals*." The conference coincided with the publication of a new French translation of *Generation of Animals* by David Lefebvre.[1] Since Allan Gotthelf was also preparing a new English translation of *Generation of Animals* and at the same time writing, together with Andrea Falcon, an article on the unity and argument of this treatise, it was decided to invite scholars who had been working on *Generation of Animals* in Europe and North America to take stock of the most significant results achieved with this important yet difficult text.

The original plan was for Allan Gotthelf and David Lefebvre to publish a volume of essays after the conference. From the very beginning, the idea was not to publish the proceedings of the conference but rather to collect in a volume some of the most interesting recent work on *Generation of Animals*, including some of the work presented at the Paris conference. When it became clear that Allan Gotthelf would not be able to complete this editorial project, the latter asked Andrea Falcon to collaborate with David Lefebvre to prepare the volume. We have selected thirteen essays, all appearing here for the first time. Two have evolved from papers presented at a fifth Pittsburgh-London (Ontario) workshop on Aristotle's *Generation of Animals* organized by Allan Gotthelf and Devin Henry at the University of Western Ontario in May 2010. One is the essay on the unity and argument of Aristotle's *Generation of Animals* that Allan Gotthelf was completing with the help of Andrea Falcon at the time of his death. This essay, which contains the most significant results of Allan Gotthelf's

[1] This translation is published in Aristotle, *Œuvres complètes* (Flammarion, Paris 2014).

lifetime philosophical engagement with this work, opens the volume. The other is by Mariska Leunissen on the order and method adopted in the second book of *Generation of Animals*. Ten essays originate from papers delivered at the Paris conference. To these essays, we have added one by Sophia Connell on Aristotle's explanation of deformity, which is a response to the most recent scholarly work on this topic.

We would like to thank Zoli Filotas who has translated Chapter 2 from French and revised the English of the Introduction and Chapter 1.

We are grateful to Michael Sharp and Hilary Gaskin for offering to publish the volume as a Cambridge Critical Guide.

The volume is dedicated to Allan Gotthelf, who was a friend, a mentor, and an example for most of the scholars involved in this project.

Abbreviations

APo	*Analytica posteriora*	*Posterior Analytics*
APr	*Analytica priora*	*Prior Analytics*
DA	*De anima*	*On the Soul*
DC	*De caelo*	*On the Heavens*
GA	*De generatione animalium*	*Generation of Animals*
GC	*De generatione et corruptione*	*Generation and Corruption*
HA	*De historia animalium*	*History of Animals*
IA	*De incessu animalium*	*On the Progression of Animals*
Juv.	*De juventute et senectute*	*On Youth and Old Age*
Long.	*De longitudine et brevitate vitae*	*On the Length and Shortness of Life*
MA	*De motu animalium*	*On the Motion of Animals*
Meta.	*Metaphysica*	*Metaphysics*
Mete.	*Meteorologica*	*Meteorology*
NE	*Ethica nicomachea*	*Nicomachean Ethics*
PA	*De partibus animalium*	*Parts of Animals*
Phys.	*Physica*	*Physics*
Pol.	*Politica*	*Politics*
Prob.	*Problemata*	*Problems*
Resp.	*De respiratione*	*On Respiration*
Somn.	*De somno et vigilia*	*On Sleep and Waking*

Introduction
Aristotle's Philosophy and the Generation of Animals

Andrea Falcon and David Lefebvre

The five books that make up Aristotle's *GA* constitute an extraordinary sum of embryological, zoological, and medical knowledge. Aristotle's overall goal is to provide a comprehensive and systematic account of how animals reproduce, including a study of their reproductive organs, what we would call fertilization, embryogenesis, and organogenesis. A few things surrounding the generation of animals are also explained in the course of the *GA*, including hybridity, birth defects, diseases of the pregnant female, and lactation. But nothing that Aristotle says in the first four books really establishes that the generative process stops at birth. Quite the opposite: *GA* 5 is concerned with the formation of parts and changes taking place in the development of the animal after its birth.[1] Moreover, a critical discussion of previous theories is part and parcel of the *GA*. Aristotle engages with the views of students of nature, doctors, and historians. He explicitly names Empedocles, Anaxagoras, Democritus, Ctesias, Herodotus, and Alcmaeon of Croton. He also makes references to doctrines and ideas that can be found in the Hippocratic treatises. By his lights, all his predecessors failed to provide adequate scientific explanations because they lacked the sort of knowledge that one acquires by means of a careful collection and in-depth study of the relevant zoological (and botanical) data. Unfortunately, how Aristotle collected and organized those data can be reconstructed only in part.[2] However, there is no doubt that the range, detail, and precision of the information collected in the *GA* presuppose a full immersion in the study of animals. Such an immersion commands our admiration.

[1] For a discussion of the investigation conducted in *GA* 5, with a focus on how it contributes to the overall aims of the *GA*, see Leunissen and Gotthelf 2010.

[2] We refer the reader to Chapter 13 in this volume for a study of the references to dissections in the *GA*.

But there is more to the *GA* than this. Aristotle is concerned with providing a robustly philosophical explanation of animal generation. By "robustly philosophical," we mean an explanation that employs key explanatory tools that Aristotle has developed for the study of nature such as matter and form, potentiality and actuality, and the doctrine of the four causes. It is telling that Aristotle opens the *GA* by announcing that we have already studied the final, formal, and material causes of each of the parts of animals except the generative ones. What remains to be done is a study of the latter and an overall study of the moving cause.[3] In this way Aristotle does not only make the moving cause a main focus of the *GA* but also gives us a fairly good idea of how his doctrine of the four causes shapes and controls a large tract of his study of animals.[4] By announcing that the moving cause will be a main focus of the *GA*, Aristotle does not mean to rule out the relevance of the other causes – especially that of the final cause. On the contrary, his study of animal generation is conducted with the crucial assumption that all natural processes, including generation, are to be studied in light of the final result and not *vice versa*. This is an important revision of the method adopted by his predecessors: while they explored what something is by studying how this thing came into existence, Aristotle studies how something comes into existence in light of what it is. Aristotle defends this revision in *PA* 1, which is a sort of prolegomenon to the study of animals, and consciously adopts it in the *GA*.[5] Also in light of this, following Allan Gotthelf, we may want to say that in the *GA* Aristotle is primarily concerned with the *formal-efficient* cause rather than the *material-efficient* cause of generation.[6]

By drawing upon explanatory resources whose significance goes emphatically beyond the study of animals, Aristotle signals his commitment to placing the study of animals in the context of a larger, philosophical project. In connection with this, it is worth recalling the famous exhortation to the study of life offered in *PA* 1.5. This exhortation insists on the overall significance of the study of animals. By Aristotle's lights, this study contributes directly to the highest form of knowledge: philosophy. Of

[3] *GA* 1.1, 715a11–14.

[4] David Lefebvre offers a detailed discussion of the opening stretch of the *GA* and how it can be used to shed light on the structure and organization of the *GA* (and beyond). See below.

[5] *PA* 1.1., 640a33–b4. For more on this important methodological insight and how it is applied in the *GA*, we refer the reader to Chapters 1 and 2 in this volume.

[6] Gotthelf 2012: 92–4. Of course, the efficient-material cause is not absent from the treatise as *GA* 5 clearly shows.

course, the philosophy Aristotle has in mind is natural philosophy. Here is the relevant passage:

> For even in the study of animals disagreeable to perception, the nature that crafted them likewise provides extraordinary pleasures to those who are able to know their causes and are by nature *philosophers*. (*PA* 1.5, 645a7–10, trans. Lennox)

We would like to elaborate on the claim that the detailed and at times almost technical knowledge achieved in the *GA* is part of a larger, philosophical plan. First of all, it is important to see that all the claims, explanations, and theories that Aristotle advances in the course of the study of animal generation not only contribute to an attempt to explain the phenomenon of life but are also an integral part of an attempt to arrive at an adequate explanation of the whole natural world. More directly, at least for Aristotle, all biological processes are natural processes – including those that are involved in the generation of an animal (and this is true even if plants and animals are the result of cultivation or raising). As such, they are controlled by a nature (φύσις) as an internal principle of motion and rest.[7] Among other things, this means that some of the explanatory principles that are introduced to explain nature apply in general to the study of biological processes and in particular to the study of the processes involved in the generation of an animal. Arguably, the most important explanatory principle is this: nature is not only matter but is also form, and it is to be studied as if we were investigating what snubness is – that is, neither independently of matter nor just in terms of matter.[8] Put differently, the generation of all animals, insofar as it involves a set of processes controlled by a nature, should be studied as we study hylomorphic compounds: neither independently of matter nor just in terms of matter.

Aristotle uses hylomorphism in the *GA* to explain not only sexual generation but also spontaneous generation. More precisely, Aristotle extends the results achieved in the study of animal generation to the study of spontaneous generation. Aristotle is forthcoming about his overall strategy at the outset of the *GA*. He tells us that he will build on the results achieved in the study of sexual generation in order arrive at a complete account of whatever types of generation there might be.[9] We should not take this strategy as a sign that Aristotle is a dogmatic thinker

[7] The most general yet precise description of nature (φύσις) as internal principle of motion and rest is offered at the outset of *Physics* 2 (192b13–17).

[8] *Physics* 2.2, 194a13–15: "οὔτ' ἄνευ ὕλης [...] οὔτε κατὰ τὴν ὕλην."

[9] GA 1.1, 716a2–4. For more on this aspect of Aristotle's strategy, see Chapter 1.

imposing a rigid theoretical framework on the complexities of the natural world. Rather, he is a resourceful explorer who is aware of the explanatory powers and adaptability of one of his most important discoveries, hylomorphism.

Of course, all this is not without tensions. What is interesting is, on the one hand, how hylomorphism imposes its own rules on the study of the generation of animals and, on the other hand, how the phenomena involved in the generation of an animal impose their own constraints on hylomorphism. For any reader of the *GA*, one obvious question is how much of the general theory of hylomorphism is found at work in the explanation of animal generation. Here we limit ourselves to pointing out that hylomorphism is characterized by the will to preserve the unity of the hylomorphic compound. Put differently, by Aristotle's lights, we do not find matter in isolation from form or form in isolation from matter in the natural world. This is what Aristotle means to say when he tells us that matter and form are separable only in thought. His chosen example, snubness, illustrates this claim in an especially vivid way. With this in mind, we may wonder how the thesis that each of the parents makes its own distinctive contribution to the process of reproduction is to be understood. According to Aristotle, the female provides the matter (ὕλη) while the male the form (εἶδος).[10] This raises the question of, on the one hand, how the matter that the female contributes to the generative process can exist in separation from the form that the male provides and, on the other, how the form that the male provides can exist in separation from the matter that the female contributes to the generative process.

Hylomorphism is introduced as a mere hypothesis at the outset of *GA* 1 but is developed into an explanatory framework for the study of animal generation in the second half of *GA* 1 (after the discussion of the reproductive parts is completed).[11] This entails finding out the distinctive contribution of the male and the female to the generative process. Aristotle approaches this investigation by asking whether both the male and the female contribute seed (σπέρμα).[12] This investigation is concluded only at

[10] Devin Henry has coined the expression "reproductive hylomorphism" to signal the specific application of hylomorphism to the study of animal reproduction. Cf. Henry 2006a and 2009: 372–4.

[11] Reproductive hylomorphism is first presented in *GA* 1.1, where Aristotle introduces the male and the female as the sources (ἀρχαί) of generation (γένεσις): the male as the source of the movement and the coming to be, and the female as the source of the matter (716a4–7).

[12] *GA* 1.17, 721b6–12. In this case, the Greek σπέρμα contains an ambiguity we have tried to capture by employing the word "seed." In the rest of the volume, we render σπέρμα as "seed" when we think that there is a latent ambiguity. When there is no such ambiguity, we translate σπέρμα with

the end of *GA* 1. Still, the separation of the two hylomorphic principles that is achieved at the end of *GA* 1 calls for an explanation. It is telling that Aristotle begins *GA* 2 with an attempt to explain why the male and the female are separate principles in most animals. What calls for an explanation is not just the separation of the male and the female as sources of generation but also their separation as sources of the form and the matter of the generative process. This separation is at the origin of two difficulties that are at the heart of Aristotle's theory of animal generation. The first is this: if the male contributes the form but no matter to the generative process, how can the form, which is by definition immaterial, act on the matter? We will not try to answer this question. Here we are content to note the special role that is reserved to vital heat and what Aristotle calls *pneuma*, conceived as an instrument or a tool employed by the form. This is one of the most interesting and original ideas Aristotle introduces in the course of the *GA*. After all, it cannot be a mere coincidence that the *GA* ends with a reflection on *pneuma*, which is described as an instrument for many things like the hammer and anvil in the art of the smith.[13] The second difficulty may be introduced as follows: if the female contributes only matter and no form, while the male provides a form that contains all the characteristic traits of the parent, including its sex, how can females be generated? This question is answered in *GA* 4.1. Aristotle explains sexual differentiation by outlining a set of physical laws that he also uses to explain how certain physical characteristics are passed on from the parents to the offspring. As we will see shortly, this explanation has attracted a great deal of attention because of its potential implications for Aristotle's theory of form.

The importance of placing the *GA* in the context of a philosophical project that goes emphatically beyond Aristotle's study of animals can hardly be overestimated. On the one hand, it is only by insisting on this larger project that the true scientific value of the enterprise attempted in the *GA* is vindicated. If we separate Aristotle's *GA* from this project, we risk reducing its contents to a sum of prejudices and empirical investigations. On the other hand, it is only in his study of animals, and in the theory of animal generation in particular, that Aristotle discusses, demonstrates, and ultimately substantiates some of the claims he makes in the

"semen." We note, in passing, that the Greek γονή, which is also used to refer to male contribution, does not have this ambiguity.

[13] *GA* 5.8, 789b2–15. Aristotle makes this comparison in the course of his criticism of Democritus and his explanation of the formation of teeth (an explanation given only in terms of efficient-material causality).

context of his natural philosophy and in the course of the project that he calls "first philosophy."

The relation between Aristotle's first philosophy and Aristotle's study of animals is a large and difficult topic. Here we are content to recall that a main focus of the project that Aristotle calls first philosophy is substance (οὐσία), and that this substance is largely conceived as a living substance. In light of this conception, we should not be surprised to see that some of the claims that are made in the course of the study of substance are based on the scientific results achieved in the study of animals. *Metaphysics* 7.7–9 is a good case in point. There, Aristotle is concerned with establishing the priority of form. However, the arguments that he uses are not fully demonstrated in this stretch of text; they are only so in the *GA*.

At the same time, it should not be lost that the explanation of animal generation advanced in the *GA* has a significance that goes beyond the boundary of Aristotle's natural philosophy. Let us focus on an especially clear example. Aristotle's theory of family resemblance has been the focus of a rich and complex discussion because of its potential implications for how we understand Aristotle's essentialism. By "Aristotle's essentialism," we mean the thesis that the form is the essence of the living substance.[14] At first, it may not be obvious that the theory of animal generation has potential implications for how we conceive of the form. Recall, however, that Aristotle is committed to hylomorphism. In other words, Aristotle explains the reproduction of an animal species as the transmission of the relevant form from the male parent to the (male or female) offspring. However, no theory of animal generation is complete without an attempt to explain how certain physical characteristics are transmitted from the male and female parents to the offspring. Aristotle's theory is no exception to the rule: it contains a notoriously difficult but very important explanation of how certain physical characteristics are passed on to the offspring. This explanation is advanced in *GA* 4.3. Given that this explanation entails the view that the male parent transmits the form but no matter, it would seem that the form that the male parent passes on to the offspring is a form that contains some of the physical characteristics that the latter inherits from the former. Hence, in a certain reading of this explanation, the form that the male parent passes on to the offspring is a robustly particular form, that is, a form that is distinct from that of all the species-members not merely in number but also because it includes, or is

[14] In Chapter 8 of this volume, Jocelyn Groisard shows that Aristotle's explanation of hybridity has important implications for his essentialism.

responsible for, some of the characteristics that distinguish the individual from the other species-member.[15]

We have offered a few remarks that we hope help to see how the theory offered in the *GA* relates to Aristotle's first philosophy. It is now time to try to situate the study of animals of which the *GA* is a part into the project that Aristotle calls natural philosophy. A brief yet precise outline of this project is offered at the outset of the *Meteorology*. There, Aristotle sketches out a very ambitious program of investigation. This program displays not only a distinct explanatory structure but also a clear organization: we begin with the study of the first principles of nature and motion. When these principles are in place, we tackle the study of the natural world by means of a study of the heavens and continue with a detailed study of the sublunary world. We approach the latter via a study of the elements and a treatment of the general principles of generation and corruption. When these additional principles are introduced, we turn to what Aristotle calls meteorology. It is only after we have completed our investigation of the meteorological phenomena that we are ready to deal with animals and plants. Here is the relevant passage:

> After we have dealt with all these subjects, let us then see if we can give some account, along the lines we have laid down, of animals and plants, both in general and separately; for when we have done this we may perhaps claim that the whole investigation which we set before ourselves at the outset has been completed. (*Mete.* 1.1, 339a5–10, translated by H. D. P. Lee, with modifications)

A full discussion of this passage goes beyond the scope of this introduction. Suffice it to say that this text is a carefully written outline of the project that we often call (for lack of a better word) Aristotle's biology. Aristotle conceives of the study of life as consisting of two separate studies, the study of animals and that of plants, to be approached in this very order: first animals and then plants.[16] Moreover, he is very clear that both studies have to be approached by deploying the explanatory principles first introduced and then adopted in the study of the other natural phenomena. This is how we think the words "along the lines we have laid down" (κατὰ τὸν ὑφηγημένον τρόπον) should be understood. This confirms the strong integration of the studies of animals and plants into a larger project that goes all the way back to the *Physics*. Of course, in order to carry out the

[15] For a review of the argument for the existence of robustly particular forms in Aristotle's philosophy, see Chapter 10 in this volume.

[16] For the significance of this order, see Falcon 2015.

study of animals we may need either additional principles or more specific principles tailored to the study of animals. The latter are introduced in *PA* 1.[17]

But how does Aristotle conceive of the study of animals? Right at the beginning of the first book of the *PA*, we are introduced to a distinction that is central to Aristotle's investigation. This is the distinction between two stages of inquiry: the first is the collection and organization of the relevant data, while the second consists in the search for their explanation.[18] The first stage is represented by the *HA*, which is concerned with laying down the differentiating features of animals. The second stage is conducted in a number of works, including the *PA* and the *GA*. The relation between these two works is clearly stated at the outset of *PA* 1, where Aristotle says that generation (γένεσις) is for the sake of being (οὐσία), and not *vice versa*.[19] We have already seen that Aristotle thinks that we have to have a grasp of what the animal is before we can successfully engage in a study of how that animal has come into existence. Now we can add that this insight entails that the study of the *PA* comes before the study of the *GA* in the order of presentation. Clearly, the investigation of animal generation comes very late in the order of study.[20] In other words, a great deal is already firmly in place as Aristotle approaches the topic of animal generation: in addition to his treatises on natural philosophy and his detailed study of the various parts of the natural world, the study of the soul as well as a large portion of the study of animals are already in place. As a direct consequence of this particular position in this order of study, the *GA* displays a number of systematic relations with works that contribute not only to the general study of nature (e.g., *Physics* 2 and *GC* 1) and to the study of aspects of the sublunary world (e.g., *Meteorology* 4) but also to the theory of the soul (*DA*) and the study of what Aristotle calls "animal nature."[21]

[17] For an introduction to the contents of *PA* 1, see Lennox 2010. Here we will not pursue the question of the relation between *Physics* 2 and *PA* 1. Suffice it to say that there is theoretical affinity between the two books. This is exactly what is expected if we take seriously the idea that the principles introduced in *Physics* 2 (hylomorphism, and the doctrine of the four causes) are further specified in *PA* 1 for the study of animals.

[18] *PA* 1.1, 640a10–12.

[19] *PA* 1.1., 640a33–b4.

[20] The order of study largely coincides with the order of explanation. This is not to be confused with the order of discovery. We know very little – in fact, next to nothing – about the latter.

[21] This phrase is used at *PA* 1.5, 645a5–6 ("ζωικὴ φύσις"). The Greek could be rendered as "living nature," but we opted for "animal nature" because Aristotle's focus is on animals, to the exclusion of plants.

So far we have been concerned with the place of the *GA* within Aristotle's larger philosophical project. We hope to have shown that an adequate study of the *GA* must take into account the larger context in which the investigation of animal generation is situated. Without this larger context, a full appreciation of the *GA* as an instance of what is by Aristotle's lights a scientific investigation of animal generation is not possible. It remains to be seen how much of the procedures that are mandated in the *APo* are at work in the *GA*. To see why this is important, we have to recall that the *APo* contains Aristotle's theory of science. If we take the *APo* to be the official presentation of Aristotle's theory of science, the obvious question is how much of this theory is found in the actual practice of science conducted in the *GA*. This is a large and difficult question. For almost forty years now, a great deal of research has been devoted to the discussion of the relations between the writings devoted to the study of animals and the *APo*. This research has focused on the zoological corpus as a case study for the application of the procedures mandated in the *APo*. We are not able to do justice to this complex debate in this introduction. We refer the reader to the contribution to this volume by Robert Bolton, who carries further an investigation he first offered in 1987. Bolton is concerned with the search for the definition, which is a crucial aspect of Aristotle's scientific enterprise. According to the theory of scientific inquiry expounded in the second book of the *APo*, this search unfolds in stages. There has been a considerable debate on how exactly we should understand this search. More directly, the number and precise nature of the stages are among the things that are debated. Bolton discusses the merits of the three-stage view of inquiry. On this view, the three stages would be: [stage 1] one gives an account of what a name or name-like expression signifies; [stage 2] one establishes the existence of the things that is signified; and [stage 3] one establishes what the thing is by arriving at its definition. Against this view, Bolton argues that the stages of inquiry can be only two: [stage 1] one establishes that something is the case and [stage 2] one establishes what this is by arriving at its definition. What is especially interesting in Bolton's approach is that he does not only use the theory of the *APo* to cast light on the investigation conducted in the first book of the *GA* but he also employs the latter to illuminate the theory offered in the second book of the *APo*.

We would like to end this introduction by turning to the internal unity of the five books that make up the *GA*, which is a central concern of the first part of this volume. On the one hand, scholars are struck by the quality of the overall construction of the treatise. Aristotle is generally attentive to the transition between books and between the major parts of his work.[22] Among other things, he is forthcoming in announcing his various investigations and situating them within his larger research project.[23] He also clearly distinguishes among various types of order that he adopts over the course of his analysis (e.g., the chronological order of generation, and the order of explanation from the most perfect animals).[24] On the other hand, scholars have noted a certain number of difficulties concerning the editorial and theoretical coherence of the treatise. For example, what is the relation between *GA* 1 and the rest of the treatise? How does *GA* 5 relate to the previous discussion?[25] Is the explanation of the hereditary differences offered in *GA* 4 consistent with the general principles formulated in *GA* 1 and 2?[26] The volume as a whole (and most obviously the essays collected in the first part) addresses these questions, both directly and indirectly, exploring different lines of interpretation.

We would like to elaborate on this last point. Darwin has popularized the use of the distinction between lumpers and splitters.[27] He originally referred to two kinds of practices of taxonomy in botany: lumpers prefer to use groups established on the basis of broad similarities, splitters tend to break apart the groups to stick to the finer differences. G.E.R. Lloyd has taken up this distinction to characterize two attitudes about the relation between the theory of science set out in the *APo* and the practice of Aristotle in the biological treatises: while the former are inclined toward reconciliation of theory and practice of science, the latter prefer to insist on what they perceive as tensions if not outright discrepancies between the two.[28] This distinction, schematic but convenient, has been taken up by Allan Gotthelf, who has reformulated it by distinguishing between integrators and fragmenters.[29] These labels can be usefully applied to the debate on the unity of the *GA*. Insofar as they are concerned with this question, the contributors to this collection are integrators rather than splitters. However, they explore different lines of interpretation and ultimately opt for different types of unity. They disagree about whether the unity of the *GA* can be stronger or weaker, theoretical or editorial, structural or linear. Thus, for Allan Gotthelf and Andrea Falcon, the treatise reveals the strong unity of a single explanatory project. Mariska

[22] E.g. *GA* 1.16, 721a26–30; 2.1, 731b18–23; 4.1, 763b20–27. [23] *GA* 1.1, 715a1–28.
[24] *GA* 2.4, 737b25–27.
[25] For an attempt to answer this question, see Leunissen and Gotthelf 2010. [26] Cf. Henry 2006b.
[27] In his letter to J. D. Hooker dated August 1, 1857 (*The Correspondence of Charles Darwin*, volume 6, 1856–57, Cambridge University Press, Cambridge, 1990: 438).
[28] Lloyd 1996: 7–9. [29] Gotthelf 2012: 185.

Leunissen, while insisting on the organized character of the treatise as a whole, concentrates on *GA* 2 to show how Aristotle harmonizes the chronological order adopted to account for embryogenesis with the observation that there are different levels of perfection and ultimately a natural hierarchy among animals. From a different perspective, David Lefebvre shows how the prologue to the *GA* aims at combining, not without difficulties, the systematic and the editorial unities of his zoological research.

The five books of the *GA* are a complex unity and the result of the integration of various investigations into a single project. While this claim is relatively uncontroversial, a perceptive reader may wonder how successful Aristotle is in assembling the various pieces that make up the five books of the *GA*. *GA* 1 is a case in point. The first two essays in this volume offer alternative interpretations of this book. On the one hand, Allan Gotthelf and Andrea Falcon see *GA* 1 as the combination of a *PA*-like investigation on the generative parts (*GA* 1.3–16) with a study of generation via a study of the male and female contributions to generation (*GA* 1.17–22). On the other hand, David Lefebvre argues that *GA* 1 is a study of the generative parts, both anhomeomerous and homeomerous (*GA* 1.3–22), integrated into a study of the principle of generation (*GA* 1.1–2 and *GA* 1.23). It is up to the reader to evaluate the merits of these alternative readings of *GA* 1. What we would like to emphasize is that the different interpretative strategies adopted in these two essays do not deny that Aristotle attempts to combine different pieces into a single project. After all, the prologue to the whole *GA* is evidence that Aristotle thinks of the *GA* as the combination two studies: a study of the generative parts and a study of the principle of generation, the moving cause.

On a different level, which does not affect the unity and coherence of the *GA* as a whole, there is evidence that materials of various provenance have been integrated into *GA* 1. Consider, in particular, the stretch of text that is printed in the Bekker edition at page 724b12–22. This passage is difficult to reconcile with its immediate surroundings. Tellingly, there is no consensus on it: while Peck considers these lines an interpolation that interrupts the argument, David Balme retains them. Drossaart Lulofs, the editor of the *GA*, prints them in square brackets. Here Aristotle defines seed (σπέρμα) as the first mixture that contains the male and the female contributions. Perhaps we can take this passage to be an attempt to answer the question left open in the preceding lines, where Aristotle wonders whether seed is either matter or form or both. What justifies the insertion of this apparently extraneous material is that it elaborates on the third possibility, namely that seed is the mixture of matter and form.[30] A full discussion of this passage would go beyond the aims of this introduction. After all, these lines present us with a local problem. Still, they provide evidence that *GA* 1 was not written all at once. Indeed, it was evidently assembled over the course of an extended research project, and in a few places we can still feel the rough edges that betray the editorial work of its author. This point can be extended to other books of the *GA* (most notably *GA* 5).

A final word on the primary goal of this volume as a whole is in order. The primary goal is to shed some light on the *GA* itself: its unity and structure, the main theses defended in the work, as well as the method of inquiry adopted. Starting from this vantage point, we hope to show how the *GA* relates to the rest of Aristotle's thought – not only his natural philosophy but also his first philosophy and theory of science. As Allan Gotthelf has shown us both in his teaching and in his writings, a full and in-depth appreciation of the project attempted in the *GA* entails an assessment of how this work relates to the rest of Aristotle's philosophy. We cannot have one without the other.

[30] Cf. Lefebvre 2016.

PART I

The Unity and Structure of Aristotle's Generation of Animals

CHAPTER I

"One Long Argument"? The Unity of Aristotle's Generation of Animals

Allan Gotthelf and Andrea Falcon

Introduction

We believe that the five books of Aristotle's *GA* are a single project, in answer to a single question, and thus have a single plan that runs from beginning to end. As a consequence, it seems to us that *GA* is a complex structure in which later parts depend on what has come before, so that it would not be effective to rearrange the pieces. Our sense is that the unity of *GA* derives from the way in which each of its parts is needed to make the case for the thesis that the entire argument is meant to support. This thesis, which represents also the conclusion of the argument, can be stated as follows: *In sexual generation, the animal offspring comes to be as it does from parents one in form with the offspring because of the transmission of form from the parents (primarily from the male) via an irreducible active potential to act upon material provided by the female (which material is such as to have a passive potential to be so acted upon) in ways that vary according to the type of animal in question (some of them live-bearing, some egg-laying, some grub-producing) – or by a process and factors analogous to that, deriving not from parents one in form with the offspring, as in spontaneous generation. And this*

Aristotle's *GA* remained at the very heart of Allan Gotthelf's philosophical engagement with Aristotle throughout his whole career – from his first article, *Aristotle's Conception of Teleology* (1976/77), where Allan offers an interpretation of Aristotle's teleology in terms of what he calls Aristotle's conception of an irreducible potential for form, to his last, *Teleology and Embryogenesis in Aristotle's GA* II 6 (2012). In Allan's mind, the article that appears here for the first time was the necessary completion, and indeed vindication, of his reading of Aristotle's *GA*. I, Andrea Falcon, worked with Allan on this article from the Fall of 2010 to his death on August 30, 2013. When it became clear to Allan that he did not have much time left to live, this article became his top priority. I was able to correspond with him until August 4, 2013. What appears here is an abridged version of a much longer article. I am confident that this shorter version captures all the essential points made in the longer article, and that it represents accurately how Allan read and understood Aristotle's *GA*. I am happy to acknowledge that not only the original impetus for this article but also most of the ideas collected in it are Allan's. I consider myself fortunate to have been his friend and a collaborator on this project. I hope to be able to publish the longer version of this article together with our (Allan's and mine) joint translation of Aristotle's *GA* in the near future.

process, in the case of sexual generation, takes place entirely because it is needed for the generation of the offspring. In the pages to follow, we will collect textual evidence to help the reader evaluate our interpretation. We will begin with the overall aim of *GA* and continue with the argument of *GA* to see how the former is realized in the latter.

The Aims of *GA*

For the statement of the overall aim of *GA*, and its setting in the larger enterprise of the study of animals, the place to start is outside *GA*, in a passage in *PA* 1.1. Having argued that coming into being (γένεσις) is for the sake of being (οὐσία), and not the reverse, Aristotle adds:

> (i) Hence we should, if possible, say that because this is what it is to be a man, therefore he has these things; for he cannot be without these parts. Failing that, we should get as near as possible to it: we should either say altogether that it cannot be otherwise, or that it is at least good thus. And these things follow. (ii) And because he is such a thing, his coming to be necessarily happens so and is such. And that is why this part comes to be first, and then this. (*PA* 1.1, 640a33–b4; trans. Balme)

In this passage, Aristotle outlines an aim for both *PA* and *GA*. While *PA* is concerned with the explanation of why animals have the parts they do, *GA* focuses on the process by which those parts come to be present. The second part of our passage elaborates on how this process has to be explained. The explanation must start from the nature of what is to be produced: it is because the product has the nature it does that certain things follow (of conditional necessity) regarding how it must come into existence. And that is why, Aristotle adds, the parts come into existence in *the order* in which they do.

What we read in *PA* 1.1 has to be integrated with what we are told in the opening paragraph of *GA*. There we are told that we have already studied the final, formal, and material causes of each of the parts of animals except for the generative ones. What remains to be done is a *PA*-like study of the generative parts and an overall study of the moving cause:

> There remain of parts the ones that contribute to generation, about which nothing was delineated earlier, and in regard to the moving cause, what the source is. To inquire about this cause and about the generation of each animal is in a way the same thing. (*GA* 1.1, 715a11–14)

According to this passage, a primary focus of *GA* is "the source of the moving cause." To put it differently, *GA* is an attempt to answer the following question: *What agent(s), and what actions of theirs, bring animals and their parts into existence?* If we bring to bear on this passage from *GA* what we read in *PA* 1.1, we may add: *And in what order? PA* 1 mentions an account in terms of conditional necessity. This brings our attention to *the final cause* of animal generation: the production, or existence, of an animal of a certain nature. A full account of the generation will not only include but also connect the final and the efficient causes of generation. It will establish the order in which the generation takes place and explain why the order is that way. The need for the teleological dimension explains why Aristotle says that an account of the moving cause is an account of generation *in a way*. With this qualification, Aristotle does not rule out that other causes, in addition to the efficient cause, are involved in the explanation of generation. In fact, it is only with the teleological dimension in place that we have a full account of the coming into being of an animal.

GA 1: Aristotle's Initial Move

With the overall aim of *GA* in place, we can turn to its argument. The latter begins with the following three-part observation: in many cases, animals come into existence from (1) the *coupling* of (2) *male* and (3) *female* (*GA* 1.1, 715a18–20). This is our initial observation – what we observe, when we first look. It is one observation because the three items that constitute it are interdefined. What we observe is a coupling, that is, the insertion by one organism of a part into the other organism. This observation originates the initial definitions of the male as that which is capable of generating into another, and of the female as that which is capable of generating into itself (*GA* 1.1, 716a13–15). So neither "male" nor "female" nor "coupling," as we observe in most cases, makes sense without the other two. Our observation is concerned with the moving cause of generation, which is to say the whence of the motion that is the coming into being of an animal offspring. In other words, the move that opens the argument of *GA* is perfectly aligned with the stated aim of the work: a study of the generation of animals largely (but not entirely) understood in terms of the moving cause.

A great deal follows from this initial move. As an immediate consequence of the choice to begin with the coming into existence of an offspring that is the product of the coupling of a male and a female, we must start our explanation of generation from the sources or principles (ἀρχαί) of that coming into being: the actual parents and their reproductive powers. Put differently, the process that we aim to explain begins with the parents, so our explanation must start from them. In particular, *it must include a discussion of their reproductive parts.* The latter are considered organs or instruments for their reproductive powers (*GA* 1.1, 716a23–25). Therefore, if we want to study the reproductive powers of the male and the female, we have no choice but to study how they are realized in the relevant bodily parts. This explanatory strategy is the reason why Aristotle turns to a study of the reproductive parts as soon as the initial definitions of the male and the female are in place. This study occupies the first part of *GA* 1 (*GA* 1.2–16).

This explanatory strategy is also the reason why the discussion of reproductive parts is postponed in *PA*.[1] Aristotle could (and would) have spoken of the reproductive parts in the context of his treatment of the parts of animals, if he did not need to speak about the coming into being of animals. But since he has to speak of the coming into being of animals, he makes a determination that it is only in the context of the latter account – which is a fuller account of the reproductive capacities and functions of the male and the female as sources or principles (ἀρχαί) of generation – that he can optimally explain why there is not only a general difference between the generative parts of male and female animals, but also further differences between the generative parts of both male and female members of different species. If we return to the opening lines of *GA*, where the overall aim of the investigation is outlined, we find that Aristotle announces an argument that brings together two things: a study of generative parts and a study of the coming into being of animals, to be approached in this very order:

> Our argument (λόγος) has brought them together, putting these parts [sc. the parts that contribute to generation] last among the things to do with parts, and the beginning of our account of generation next after that. (*GA* 1.1, 716a2–4)

Another consequence of the initial move, and the ensuing focus on sexual generation, is the need to understand the respective contributions to the generative process of the male and the female, including the extent to which each contributes seed (σπέρμα). It is telling that, already in the opening chapter of *GA*, the study of σπέρμα is predicted to be of the utmost importance in understanding the distinctive roles of the male and the female in the coming into being of an offspring:

> In accordance with what we said, one might posit the male and the female as sources of generation [. . .] one would most of all come to believe

[1] Cf. *PA* 4.4, 678a21–26.

> this if one studied how the seed (σπέρμα) is produced and from where. (*GA* 1.1, 716a4–8)

Also in light of this anticipation, it is no surprise to find out that Aristotle turns to the study of σπέρμα as soon as he has completed his study of the reproductive parts. Here is what Aristotle says as he introduces the topic of σπέρμα in *GA* 1.17:

> It is thought that all [animals] come from seed, and that the seed comes from those that generate. Therefore, it is part of the same argument (λόγος) [to establish] whether both male and female emit seed and whether it comes from all the body or not from all; for it is reasonable, if it does not come from all the body, that it may not come from both the generators either. Since some say that it comes from all the body, we must first examine how the matter stands. (*GA* 1.17, 721b6–12)

The question of whether the σπέρμα comes from both the male and the female is linked to the question of whether it comes from all of the body. If not the latter, then the former need not be true either. This link is reaffirmed at the end of the discussion of pangenesis as a way to introduce the discussion of the respective contributions of the male and the female, beginning with the role played by the female:

> If the female does not emit seed, by the same argument neither does it come away from all [body], and if it does not come away from all [body], there is nothing absurd in the seed not coming from the female, but the female being a cause of generation in some other manner, which is what is to be investigated next, since it is apparent that the seed is not separated off from all the parts. (*GA* 1.17, 724a7–13)

The question of whether the female emits σπέρμα is the entry point into the larger question of how the male and the female contribute to the generative process. The latter is the question that concerns Aristotle in the second part of *GA* 1 (*GA* 1.17–23). This stretch of text provides *the basic framework* of an answer to this question, *a schema* for understanding the role of the male and the female *qua* contributors to the generation of the offspring. We will have to see how this schema is filled out in the rest of *GA*. For the time being, let us concentrate on the progression of the argument in *GA* 1.17–23. Aristotle is quite forthcoming about his argumentative strategy: the starting point of his investigation is to establish what σπέρμα is (*GA* 1.18, 724a14–15). It is neither a part of the body, nor some unnatural growth, nor nutriment taken in from outside, nor a colliquation, and so it must be a residue. It must be a residue of the useful nutriment at its final stage – blood concocted and somehow particularized (*GA* 1.18, 726a26–28 combined with *GA* 1.19, 726b5–6). As soon as he has established what σπέρμα is, Aristotle turns to the female contribution to generation in order to establish what it is. He is able to establish that the blood-like secretion produced in the female (the menses or καταμήνια) is a useful residue, having its nature because of the colder (and thus weaker in ability to concoct) character of the female (*GA* 1.19, 726b30–727a2). This contribution is not σπέρμα: two spermatic secretions cannot be produced at once; if the female produced σπέρμα, it would not have καταμήνια; but it has καταμήνια, so it cannot produce σπέρμα (*GA* 1.19, 727a25–30). He is also able to establish the nature of the female contribution to the generative process: *the female contributes the matter* (*GA* 1.19, 727b32). At this point, Aristotle turns to the male contribution to establish, again, what it is. He approaches this question by presenting two options:

> At the same time from these things it is apparent in regard to those things which are to be investigated next, how in the world the male contributes to generation, and how the male's semen is a cause of that which comes to be, whether [a] as present within and being immediately a part of the body of that which is coming to be, mixing with the matter from the female, or [b] the body of the semen does not participate but the potential (δύναμις) and movement (κίνησις) in it does, for this potential is that which acts while that which is constituted and takes the shape is the remainder of the residue in the female. (*GA* 1.21, 729b1–8)

It turns out that the latter is the position endorsed by Aristotle: the male does not contribute any matter but rather an active potential (δύναμις), which is based in certain movements (κινήσεις) conveyed by the semen (σπέρμα). A general reflection on what comes into existence from the action of an agent on a patient suggests that this is all that is needed. Furthermore, what we observe in the case of certain insects, where the male does not insert a part of itself into the female but, on the contrary, the female inserts a part of itself into the male, and the heat and the δύναμις of the male (such as it is) does the causal work, confirms this suggestion (*GA* 1.21, 729b22–33).

At this point, Aristotle has delineated an answer to the question of how the male and the female contribute to the generation of an animal. He is in a position to explain why the female parent does not produce offspring by herself: generation needs a source of motion, which is to say something that initiates the process (*GA* 1.21, 730a29–30). He is also in a position to explain why the formation and development of the embryo takes place in the female: the matter out of which the embryo comes into existence is provided by the female parent – both that out of which the embryo is formed and the additional matter supplied as the embryo grows (*GA* 1.22, 730b1–4). Aristotle uses the craft analogy to argue that the male contribute *the form* by means of his role as source of motion. In particular, in those animals that emit semen (σπέρμα), nature uses the latter as a tool with motion in actuality just as an artist uses specific tools to produce a certain outcome (*GA* 1.22, 730b19–23).

With an answer to the question of how the male and the female contribute to the generative process, the argument has come to a natural pause. Before continuing our review of the argument of *GA*, we would like to take stock of what we have learned so far:

(1) The key to the unity of *GA* 1 is Aristotle's initial move (our initial observation). This move helps dispel the special problems posed by *GA* 1 for those who seek unity in *GA* – unity not only between books but also across a book.
(2) Aristotle uses the term "λόγος" to describe what he is offering in *GA*. With reference to the study of the moving causes and the *PA*-like study of the generative parts, he says that *"our* λόγος *has brought them together"* (*GA* 1.1, 715a15). In postponing the study of spontaneous generation, Aristotle announces that the study of each kind of animals is to be conducted *"according to the ongoing* λόγος*"* (*GA* 1.1, 716a3). In introducing his discussion of σπέρμα, Aristotle tells us that *"it is part of the same* λόγος *whether both male and female emit seed and whether it comes from all body"* (*GA* 1.17, 721b7). In all three cases, we have translated λόγος with "argument" rather than "account" or "exposition." As we proceed in our analysis of *GA*, it will become clear that the λόγος in question is a logical narrative in which subsequent components build on previous components, and that together they form a unity leading up to a single complicated thesis – the conclusion of the argument we have expressed in a single statement at the outset of our essay.
(3) *GA* 1 is the first installment in a long argument. The book provides the beginning of an answer to our central question: *what agent(s), and what actions of theirs, bring animals and their parts into existence? And in what order?* *GA* 1 is manifestly not meant to stand on its own. A few times, the text points ahead: when it promises a discussion of milk (*GA* 1.16, 721a26–30); when it says that mutilations and resemblances are to be studied together (*GA* 1.17, 724a3–7); when it postpones dealing with the significance of multiple births (*GA* 1.20, 729a11–14.); and finally when it announces a study of the generation of the hard-shelled animals (*GA* 1.23, 731b13–14).

GA 2: A Fresh Start? Not Quite

Let us recall the observation that marks the beginning of the argument in *GA* 1: in many cases, animals come into existence from (1) the *coupling* of (2) *male* and (3) *female*. It is on the basis of this initial observation that Aristotle posits the male and the female as the sources of the coming into being of animals. Given the centrality of the male and the female in Aristotle's theory of animal generation, the argument of *GA* cannot be completed without an attempt to answer the question of why there are males and females. Furthermore, *GA* 1 ends with the claim that the male and the female are together in plants but are separate in animals (*GA* 1.23). At that point, the perceptive reader may wonder why the sources of generation have to be separate in animals if they are not so in plants. With this in mind, let us turn to the beginning of *GA* 2:

> It was stated earlier that the female and the male are sources of the coming to be, and what is their capacity and the definition of their being. Why the one comes to be and is female, the other male, insofar as it is of necessity and from the proximate mover and a certain sort of matter, our argument (λόγος) must try to tell as it proceeds; but insofar as it is due to the better and the cause that is for the sake of something, it has its source from above. (*GA* 2.1, 731b18–24)

While the first sentence makes contact with *GA* 1, the second promises an answer in two parts to the question of why there are males and females. The second part of the answer is an explanation of why it is better that the male and the female are separate principles. This explanation is given in *GA* 2.1. This is not yet an explanation of the coming into being of the male and the female. For that, we have to wait until *GA* 4.

As soon as Aristotle has an answer to the question of why it is better that the male and the female are separate sources, he makes an observation that shapes his discussion throughout *GA* 2–3. He observes that there is a great deal of variation in the degree of completeness of the immediate outcome of γένεσις: some animals produce similar, complete, live young, while others give birth to something that is not articulated: eggs in the case of blooded animals, grubs in the case of bloodless insects. Aristotle traces this variation in the degree of completeness at birth back to the nature of the generating parents: the hotter and the moister their nature is, the more complete their product is at birth (*GA* 2.1, 732b27–32). This observation is our second observation. While the three-part observation that opens the argument of *GA* controls the whole argument of *GA*, the second observation shapes the argument in *GA* 2 and *GA* 3. As we progress in our explanation of the coming into being (γένεσις) of animals, we have to take into account what is *specific* about each of the different modes of reproduction, starting from the live-bearing animals, continuing with the egg-laying animals, and ending with those that produce grubs. The study of the live-bearing animals is offered in *GA* 2. The other two kinds of animals are discussed in *GA* 3.

Before turning to the study of the live-bearers, Aristotle deals with three concatenated puzzles whose significance is general. At least two things should be noted in connection with these puzzles. First, they concern *all* modes of generation, so they have to be discussed *before* engaging in a study of what is specific about the different types of animals. Second, these puzzles naturally arise from the results reached in *GA* 1, so their discussion represents a natural development, and indeed an inevitable continuation, of the theory advanced in *GA* I. We have suggested that the second part of *GA* 1 (*GA* 1.17–23) provides a *schema* to understand the role of the male and the female contribution to the γένεσις of animals. In *GA* 2.1–3, Aristotle fills that initial schema in various, important ways. The first puzzle, which is emphatically described as a very great one, is introduced by this question: How does the plant or any of the animals come from the seed?[2] The puzzle is generated by the following considerations. On the one hand, the agent must be in contact with the product, so the male parent cannot be responsible for the internal conception, and indeed the development, of the embryo. On the other, the agent must be in actuality what comes into existence, so the agent cannot be in the semen. To solve the puzzle, Aristotle introduces the idea that contact can be transmitted in such a way that the actual parent can be responsible, as source of motion, for the final product of γένεσις even if it does not touch it. The example used to illustrate this idea is that of the "marvelous αὐτόματα" in which a single, minute movement at the beginning can initiate a complex, structured result. In the case of the "marvelous αὐτόματα," the result is perhaps a dance; in biological generation, it is the coming into existence of an animal and its parts. With the help of the illustration, we can see how a single causal power or potential transmitted from the actual parent via the σπέρμα remains active throughout the entire process, shaping and controlling it from beginning to end.

As soon as Aristotle has worked his way through the first puzzle, he is ready to turn to the question of what kind of material nature the σπέρμα must have in order to be able to carry the generative δύναμις:

[2] *GA* 2.1, 734b22–24.

> So we have said, then, what is the cause as the source of – as what primarily moves and fashions – the generation of each thing. But one might puzzle over the [material] nature of the semen (σπέρμα). (*GA* 2.2, 735a27–30)

This puzzle occupies Aristotle in *GA* 2.2. In a long and careful discussion, which makes use of the results achieved in *Meteorology* 4, Aristotle infers from the reaction of σπέρμα to heat and cold that its material nature is a combination of water and πνεῦμα, hot air (*GA* 2.2, 735b37–736a1). But what happens to this material nature, once it has conveyed the generative δύναμις of the male parent? This too is a question that arises directly from *GA* 1. Recall that Aristotle has already argued that the male parent does not contribute any material to the generation (*GA* 1.21). Right after discussing the question of the material constitution of the σπέρμα, he returns to this question via the discussion of another puzzle. This is our third puzzle:

> The next puzzle to be stated and solved is this: if, in the case of those [animals] that emit semen into the female, that which enters is no part the embryo produced, *where is its bodily part diverted, if it is true that it works through the potential that is within it?* (*GA* 2.3, 736a24–27)

Aristotle's answer to this question is twofold. He argues, on the one hand, that the bodily part of the σπέρμα dissolves and evaporates and, on the other hand, that the very thing that makes the semen fertile, the so-called heat in the πνεῦμα – which is not the heat of fire but is similar to the generative heat coming from the sun – conveys the potential for the soul. Aristotle completes his discussion of the male contribution to generation by explaining how the σπέρμα receives the requisite movements (κινήσεις) from the male parent. Aristotle has already argued that the σπέρμα is a residue of food. He adds that σπέρμα is a residue of food when food has been given the male parent's growth movements. This means that when the σπέρμα comes into contact with the καταμήνια, it imparts to the latter these very movements. In other words, the movements that are responsible for the growth of the male parent are those that are involved in the formation and development of the embryo.

Throughout *GA* 2.1–3 the focus remains on the male contribution to generation. By working within the general framework delineated in the second part of *GA* 1, Aristotle analyzes in greater detail the nature of the male contribution. By the end of *GA* 2.3, we have in place (a) a mechanism for how the generative δύναμις is transmitted via certain movements from the male parent, (b) an explanation of how the latter can operate from a distance by means of the σπέρμα as instrument conveying those movements, and (c) an account of the material constitution of the σπέρμα.

Finally, from *GA* 2.3, we learn that (d) the δύναμις transmitted through the σπέρμα is not only distinct from the causal power of the hot and the cold; it is also irreducible to them. Put differently, this δύναμις is an irreducible potential for form.

The next task that Aristotle sets for himself is to attempt an explanation of how to get from the formation of the embryo to the immediate outcome at birth. Since Aristotle has observed that there is great variation in the degree of completeness of this outcome, he has no choice but to take into account what is specific to the different modes of reproduction. But how should we proceed in the study of the different modes of reproduction? Where should we start and, more importantly, why? Aristotle tells us that:

> We must start first from what is first. The complete animals are first, and such are the live-bearing ones, and first among them is the human being. (*GA* 2.4, 737b25–27)

The sentence is crafted in a way that leaves no doubt that we must begin with the live-bearing animals because they are complete, and in particular with the human being because the latter is the most complete among the live-bearing animals. The idea seems to be that the maximum possible (*sc.* the most complex) outcome, namely the production of a human being, is the natural place to start, so we *must* start from there.

GA 2.4–6 is a study of live-bearing animals via a study of the human being. Aristotle starts from the results achieved in the general study of generation (*GA* 1–2.3). He applies those general results to the particular case of the live-bearers. The structure of the explanation remains the same as before: it moves from the actual parents and their respective roles in *genesis* to the formation of the embryo, and at least in principle to everything that follows from there. *GA* 2.4 begins with the male and the female contributions as residues of the nutritive process (with a concentration on the female residue and how it is formed). It continues with the formation of the embryo and, finally, turns to the question of its development and differentiation (διάκρισις). Aristotle offers an outline of the process of διάκρισις: the heart is the first part to be differentiated, and this is clear to the senses and in theory (*GA* 2.4, 740a1–24). He adds that the formation of the parts does not happen, as some think, because like is carried toward like. Rather, the female residue has the potential to be acted upon by the male residue, so whenever they come into contact the one acts and the other is acted upon *right away.* In other words, to explain a complex process such as the differentiation of the parts nothing else is required except the contact of the active and the passive potentials provided, respectively, by the male and the female. At this point, Aristotle invokes the analogy with craft to add one important element to this explanation: just as in the case of a craft, the product is the result of the agency of the craft by means of its instruments – or rather by means of the *movement* (κίνησις) of those instruments – so in the case of an animal, the δύναμις of the nutritive soul, "*using the hot and the cold as its instruments*," is responsible not only for the growth of the animal after birth but also for the formation and development of the embryo (*GA* 2.4, 740b25–34). In this stretch of text, Aristotle is not content to restate that the generative and the nutritive κινήσεις are two descriptions of one and the same causal power; he is also careful to distinguish this causal power from that of the hot and the cold. The hot and the cold are equated to tools or instruments used by the δύναμις of the nutritive soul. Not only does this confirm the results reached in *GA* 2.1–3; it also elaborates on the solution regarding the existence of an irreducible potential for form by applying it to the explanation of the διάκρισις of the parts. In other words, *GA* 2.4–6 builds on, and develops further, the results reached in *GA* 2.3.

GA 2.5 deals with a puzzle that arises directly from the outlined explanation of how the διάκρισις of the parts takes place. If the female has the nutritive soul and the residue provided by the female is the matter, why can the female not produce an offspring by herself? In other words, why is the male needed at all? The puzzle, Aristotle says, has some reason especially if we consider that female birds can produce unfertilized eggs that have some life in them (so-called wind-eggs). Yet, Aristotle adds, there is no διάκρισις in the unfertilized egg. On the basis of this observation, Aristotle concludes that the διάκρισις in the fertilized eggs is due to the exclusive contribution of the male parent. He specifies further the nature of the male contribution: while the female parent contributes the nutritive soul, the male parent supplies the perceptive soul. More precisely, the potential for the perceptive soul is transmitted via the σπέρμα, and when the matter is also present – just as in the case of the "marvelous αὐτόματα" – *the διάκρισις of the embryo unfolds in a determinate order* (*GA* 2.5, 741b7–9).

Order is the primary focus of *GA* 2.6. Aristotle attempts a full explanation of the sequence in which the various parts of the animal come into existence. Three things are to be noted in connection with this explanation. First, it partly satisfies the aims outlined for *GA* in *PA* 1. Similarly, it partly responds to the original questions that *GA* is expected to answer: *What agent(s), and what actions of theirs, bring animals and their parts into existence? And in what order?* Second, the question of the order in which the various parts of the animal come into existence is first introduced in *GA* 2.4. But it is only after the male contribution is fully specified in *GA* 2.5 that Aristotle is able to attempt a full answer to this question. Third, this attempt is introduced by recalling the analogy of the "marvelous αὐτόματα." This analogy conveys the message that the formation and development of the embryo are to be conceived as a single, unified process controlled by a single causal power: the δύναμις transmitted from the male parent via the σπέρμα. The account of this process given in *GA* 2.6 adds important details to what has been established in *GA* 2.1–3 and further developed in *GA* 2.4–5. There is no attempt in *GA* 2.6 to explain the formation of the parts – either all of them or just some (the uniform ones) – solely in terms of the material agency of the hot and the cold. Rather, the irreducible potential for form supplied by the male parent remains the primary causal factor throughout the entire process of differentiation. It employs the hot and the cold in the formation of both the uniform and the non-uniform parts.[3]

But how does the actual διάκρισις of the parts take place and why? According to the account offered in *GA* 2.6, right after the heart and the blood vessels that extend from the heart, the upper half of the body appears in outline. The head is formed first, and the brain right after the heart to compensate for the heat of the latter. The eyes come to be early but are completed only at a later stage due to the particular nature of their material constitution. The formation of flesh and the other sense-organs is made using the purest material. What remains is earthy and so is used to produce bones and sinews. The bones are differentiated together with the most important internal parts. Nails and hair (but also hoofs, horns, and the spurs of cocks) are formed out of nourishment taken in after birth and not used up for the growth of the other parts. This explains why they, unlike bones, keep growing throughout a lifetime and even after death. Teeth pose a special difficulty: while they are bone-like, they are formed after birth. Note that a discussion of the reason why some fall out and grow again is postponed: "*the cause why some teeth grow and fall out while others don't will be said later*" (*GA* 2.6, 745b13–15).

GA 3

GA 1 and *GA* 2 are continuous, yet *GA* 2 takes on questions that arise directly out of the project to which *GA* 1 has begun to give an answer. The observation that there is variation in offspring, made at the beginning

[3] For an in-depth discussion of *GA* 2.6, see Gotthelf 2012.

of *GA* 2, provides the framework for a study of what is specific to the different modes of reproduction. This study enriches our understanding of the theoretical framework first introduced in *GA* I and further developed in *GA* 2.1–3. The result is a progressive articulation of our initial schema. The opening lines of *GA* 3 make contact with what has been done in the second part of *GA* 2; they also announce what is next in the order of explanation:

> Now, we have spoken about the infertility of mules and about the animals that bear live young, both outside and inside themselves.[4] As for those blooded animals that produce eggs, in one way the things to do with generation are similar in them and in the land animals, and it is possible to seize upon something that is the same about them all; but in another way they have differences both among themselves and in relation to the land animals. (*GA* 3.1, 749b10–15)

To appreciate what Aristotle tells us in this passage, we have to return to the beginning of *GA* 2. After noting that some animals bear live young, while others lay eggs, and yet others produce grubs, he adds:

> All the animals that bear young or lay eggs are blooded animals, and all the blooded animals either bear live young or lay eggs, if they are not infertile. (*GA* 2.1, 738b8–9)

If we combine what we are told at the outset of *GA* 3 with this observation, we obtain the following explanatory strategy: as we proceed in our argument, we must keep in mind that certain facts about the generation of the egg-laying animals are not specific to this kind of animal but are shared with the live-bearing animals. These facts remain outside the scope of our investigation because they can, and indeed should, be explained in common for both live-bearing and egg-laying animals *qua* blooded animals. Our discussion will focus only on the causes of those facts that *distinguish* the animals that produce eggs from those that bear live young. In doing so, however, we must not overlook the differences that exist among the egg-laying animals, so our discussion will also deal with the causes of what *distinguishes* the animals that lay eggs from each other.

The study of the egg-laying animals is conducted in *GA* 3.1–8. Aristotle adopts the methodological principle spelled out in *GA* 2.4 ("*first what is first*"): he begins with the animals that lay a complete, hard-shelled egg and

[4] The discussion of live-bearing animals ends with the discussion of a special problem: the infertility of mules (*GA* 2.8). For the significance of this discussion for the argument of *GA*, we refer the reader to Chapter 8 in this volume.

continues with those that lay an incomplete egg. The overall explanatory strategy remains the same as in *GA* 1 and *GA* 2. Since we are concerned with a single causal process that can be traced back to the actual parents, our explanation must start from the actual parents. What is specific to this mode of reproduction is that an egg is produced, so our explanation must include an explanation of the egg.[5] The topic of egg-laying animals ends with a discussion of what is specific about the bloodless animals that lay eggs. The soft-bodied animals (sepias and the like) and the soft-shelled animals (*caraboi* and creatures akin to them) are discussed in *GA* 3.8. *GA* 3.9 deals with insects. Most insects produce grubs. What is distinctive about this mode of reproduction is discussed in *GA* 3.9. The generation of bees poses a special difficulty. With the discussion of this difficulty in *GA* 3.10, the explanation of sexual generation comes to an end. At this point, we have both a general account of how the male and the female contribute to generation (*GA* 1–2.3) and a detailed study of what is specific about the different modes of sexual reproduction (*GA* 2.4–3.10). We are in a position to turn to the study of the animals whose generation is not from a male and a female but is spontaneous (αὐτόματος). Along with a few of the insects, this study will attend to the generation of the hard-shelled animals and other stationary animals considered close to plants. Their study is conducted on the basis of the theoretical framework developed in the course of the study of sexual generation. Aristotle is quite explicit regarding this aspect of his strategy at the outset of his argument:

> Concerning the generation of the other animals [sc. animals that do not reproduce sexually], *we must speak about each of them according to the ongoing argument* (λόγος), *building it from what has been said.* (*GA* 1.1, 716a2–4)

This is exactly how the argument proceeds in *GA*: it begins with a study of sexual generation and ends with a discussion of the generation of those animals whose coming into existence is αὐτόματος. This discussion is postponed until the very end of *GA* 3. Spontaneous generation is envisioned as a two-stage process analogous to sexual reproduction. In the latter case, we have a two-stage process in which, first, the material principle is formed in the female and, then, is acted upon by the male principle. In spontaneous cases, the heat of the season in the surrounding

[5] Note that this explanation is given for *the complete egg*. This too is a direct consequence of the adoption of the explanatory principle "*first what is first.*" If there are differences that distinguish the incomplete from the complete egg, they will be discussed in connection with the study of the animals that lay incomplete eggs.

environment, by a process of concoction, produces something like a female residue. In rotting earth or earthy water we find πνεῦμα, and in πνεῦμα we find vital heat. Whenever some of the latter gets enclosed, it quickly leads to the generation of the animal. The language used to describe this process suggests that the active process is envisioned as an irreducible potential analogous to the immediate moving cause in the process of sexual reproduction. However, unlike the δύναμις transmitted by the male contributor, the vital heat is a non-species-specific potential for life. This means that differences in the final outcome do not depend on differences in the nature of this potential (in "amount or strength of the vital heat"). Rather, they depend on differences in the particular nature of the enclosing material. In the case of the hard-shelled animals, which are produced in the sea, the earthy water of the sea becomes the shell as the consequence of a process of solidification similar to the one that occurs in the case of bodily materials such as bones and horns.[6]

How Do *GA* 4 and *GA* 5 Contribute to the Argument of *GA*?

We can approach this question by reflecting on the opening sentence of *GA* 4:

> It was stated concerning the generation of animals, [and stated] both in common and separately concerning them all. (*GA* 4.1, 763b20–21)

This is a brief yet precise summary of the argument offered in *GA* 1–3. Aristotle has concerned himself with the coming into being of animals first *in common* and then *separately*.[7] He has combined an account that captures salient features of the process of generation common to all animals (*GA* 1.1–2.3) with a treatment of what is specific about the coming into being of each kind of animal (*GA* 2.4–3.11). The rationale for this explanatory strategy is found in *PA* 1. There, Aristotle says that the explanation of certain features of animal life should be given across different species of animals. He adds that such explanations are needed to avoid tedious repetitions. His examples are sleep, respiration, growth, decline, and death.[8] Avoiding needless repetition cannot be the sole

[6] For a defense of this reading, we refer the reader to Gotthelf 1989.

[7] It might be helpful to recall how Aristotle sums up the contents of *PA* 2–4 at the outset of *GA*: "It was stated about the other parts in animals, both in common and separately for each kind concerning their distinctive features, in what way each is on account of this sort of cause" (*GA* 1.1, 715a1–2).

[8] *PA* 1.1, 639a19–20.

motivation for the explanatory strategy envisioned in *PA* 1. The theory of scientific explanation outlined in *Posterior Analytics* requires that explanations be given at the proper level of generality. This requirement is central to Aristotle's explanatory project. *GA* is no exception to the rule. It is only by giving explanations at the right level of generality that we can capture salient features which might be otherwise missed. But this only makes our original question more pressing: once both the account that deals with what is common to all animals and the treatment of what is specific about each kind are in place, and we are able to integrate them into a single argument, what else is needed to complete the original plan of *GA*? A possible answer to this question is negative: nothing else is needed. No matter how interesting and insightful *GA* 4 and *GA* 5 are, these books are mere add-ons to the argument of *GA*. The latter is complete at the end of *GA* 3. We reject this conclusion. We believe that *GA* 4 and *GA* 5 are an integral part of the original plan of *GA*. By our lights, these books contribute, immediately and directly, to the main argument of *GA*. More to the point: without them, the argument offered in *GA* would, *in Aristotle's eyes*, be incomplete.

GA *4*

Let us return to the beginning of *GA* 2:

> It was stated earlier that the female and the male are sources of generation, and what is their capacity and the definition of their being. *Why the one comes to be and is female, the other male, insofar as it is of necessity and from the proximate mover and a certain sort of matter, our argument (λόγος) must try to tell as it proceeds*; but insofar as it is due to the better and the cause that is for the sake of something, it has its source from above. (*GA* 2.1, 731b18–24)

The promise made at the outset of *GA* 2 is not fulfilled in *GA* 2–3. It is only at the beginning of *GA* 4 that Aristotle attempts an explanation of why some animals become males and others become females. Right after having recalled in outline the argument of *GA* 1–3, Aristotle comes back to this question:

> But since in the most complete of them the female and the male are separated, and since we assert that these capacities are sources in all animals and plants alike [. . .] it is to be spoken first about the coming to be of these [sc. the male and the female]. (*GA* 4.1, 763a21–26)

What we need to understand is why this question cannot be answered in *GA* 2.1. One possibility is that the answer depends on explanatory principles not yet available at the outset of *GA* 2. Another possibility is that the question of why some animals become males and others females is part of a set of questions that are to be set off from the questions dealt with in *GA* 2–3. In this case, it would be a fact about subject matter, rather than a fact about the answer given or the explanatory principles used in giving the answer, that would explain why an attempt to answer this question is postponed until the beginning of *GA* 4. Both considerations are relevant.

In regard to the first consideration: the issue of why some animals become males and others females requires an analysis in terms of *the material-efficient cause*. This is already clear from what Aristotle says at the outset of *GA* 2: "*it is of necessity and from the proximate mover and a certain sort of matter*" (GA 2.1, 731b21–22). But we have seen that the focus of *GA* 2.1–3 is on the male contribution to generation. There, Aristotle offers an analysis of the male contribution in terms of *the formal-efficient cause*. This type of cause remains a primary focus as Aristotle proceeds to a separate study of different modes of animal reproduction throughout *GA* 2 and *GA* 3. It is only when that study is completed that Aristotle is able to return to the question introduced in *GA* 2.1. In regard to the second consideration: when Aristotle returns to the question introduced in *GA* 2.1, he does not say that we must speak about the *genesis* of the male and the female; rather, he says that we must speak *first* about the *genesis* of the male and the female. Evidently, he has a larger agenda in mind in which the answer to the question of how the sex of the offspring is determined leads to the explanation of something else. To learn more about this agenda, we have to turn to the beginning of *GA* 4.3. There, Aristotle tells us that the causal mechanism that explains why some animals come to be males and others females explains also the resemblance (and non-resemblance) between offspring and parents (and other ancestors). In brief, Aristotle employs his explanation of the coming to be of the sexes as an introduction to the causal mechanism that explains inheritance.

By now, it should be clear why the issues discussed in *GA* 4.1–3 are taken up at the beginning of *GA* 4 and not earlier. A rearrangement of the argument of *GA* might have been possible, but it would not have been effective. Aristotle could have attempted an answer to the question why some animals becomes males and others females in *GA* 2.1. But that would have entailed separating the question of sexual determination from the question of family resemblance. Given that Aristotle thinks that these two questions are causally connected, he opts to discuss them together. It should also be clear that *GA* 4.1–3 is an integral part of the argument of *GA*. GA 4.1–3 completes the account of generation given in *GA* 1–3.

GA 5

A primary focus of the investigation conducted in *GA* is the causal process leading to the birth of the offspring. But nothing that Aristotle says in the first four books establishes that the generative process stops at birth. Quite the opposite: if we take seriously the claim that a final-causal account has to be integrated with an efficient-causal account, as stated in *PA* 1.1, where Aristotle outlines an aim for *GA* by saying that the nature of what is to be generated is the starting point for an explanation of generation, the subject matter of *GA* must be *the coming into being of the mature organism.*

We have recalled this passage from *PA* 1.1, and its relevance to the argument of *GA*, because we need an adequate concept of coming into being (γένεσις) if we want to understand how *GA* 5 contributes to the overall argument of *GA*. We would like to suggest that GA *is concerned with the coming into being of an animal in the sense of the natural, continuous process from a fully developed animal to another fully developed animal of the same kind.* As a result of this, the study of animal generation cannot stop at the birth of the offspring but must continue with a logical focus on the mature organism. The focus on the mature living being does not preclude an interest in some changes past the acme of the animal. In the course of *GA* 5, Aristotle is able to explain changes that take place in old age such as growing grey or growing bold. This happens whenever the same causal mechanism that explains why certain features are acquired in the course of maturation is still at work in old age and can be used to explain what happens to the aging organism. With these considerations in place, we can turn to the opening lines of *GA* 5:

> We must now study the attributes by which the parts of animals differ. I mean such attributes of parts as blueness or blackness of eyes, and height and depth of voice, and differences of color in hairs and feathers. (*GA* 5.1, 778a16–20)

While *GA* 4 concludes with a discussion of phenomena related to the growth of animals within the womb and to their birth, *GA* 5 continues with a study of the attributes by which the parts of animals differ. The attributes singled out in this passage are the result of changes that occur in the course of the development of the offspring *after its birth.* They are not realized uniformly in all animals. Rather, for some kinds of animals these attributes are possessed universally, whereas in others, and particularly in human beings, the attributes are variable and occur randomly. This explains why *GA* 5 has a special focus on human beings: these attributes are most visible, and sometimes *only* visible, in human beings. Their dominant (but not necessarily only) mode of explanation is a non-teleological one appealing primarily, or at least first, to material-efficient causes.[9]

The focus on attributes that are not universally held by all animals and are primarily explained by material (rather than conditional) necessity goes some way toward explaining why the attributes of the parts of animals discussed in *GA* 5 are not treated in *PA* 2–4. The project of *PA* 2–4 is fundamentally that of establishing *for the sake of what* the parts of animals exist. Hence, the dominant (but emphatically not exclusive) mode of explanation in *PA* 2–4 is a teleological one appealing to formal-final causes. In this explanatory context, differences that are not uniformly realized within a whole species, or that change over the animal's lifetime, cannot appear as explananda.[10] But a reason why the discussion of certain attributes does not pertain to the project of *PA* 2–4 is not a reason why it pertains to the project of *GA*. We still have to find out the reason why the discussion of these attributes is part of the explanatory agenda of *GA*. An answer is easily at hand if we accept that generation is a process beginning with a fully developed animal and leading to another fully developed animal of the same kind. In particular, if the generative process does not stop at birth, then the explanation of certain facts that belong to the later development of the animal is to be expected from the theory of generation. More to the point: without an explanation of those facts, our theory of generation would be incomplete.

[9] For an in-depth discussion of the style of explanation adopted in *GA* 5, see Leunissen and Gotthelf 2010.

[10] They cannot appear as explananda because they cannot be explained from the essential nature of the organism.

CHAPTER 2

Parts and Generation
The Prologue to the Generation of Animals *and the Structure of the Treatise*

David Lefebvre

Introduction

The structure of the *GA* has sometimes been thought to detract from its scientific worth and indeed its intelligibility. In his *History of Embryology* – a reference work published in 1934 and again in 1959 – Needham praises the treatise for offering "the first great compendium of embryology," but complains that its organization is haphazard and that it is replete with digressions, repetitions, and other signs of the contingencies of its production.[1] *GA* 5, which was once believed not to belong to the treatise at all and is sometimes still treated as an appendix, is an extreme example of the problems with the work.[2] The prologue at the beginning of *GA* 1 is likewise sometimes considered a later addition: it is bracketed in Balme's edition.[3] As with other Aristotelian treatises, then, the work's structure calls for preliminary study, lest it interfere with further interpretation.[4] And although Aristotle's presentation may derail audiences used to modern scientific works, this does not mean that the *GA* is a collection of disparate essays or lecture notes. In fact, the treatise is rigorously and in certain respects systematically constructed. And it is all the more important to shed light on its structure because that structure is rooted in Aristotle's epistemology and metaphysics, and Aristotle takes the scientific value of the study to depend on it.

Thanks especially to work by Jim Lennox, Allan Gotthelf, and Pierre Pellegrin, it is now widely recognized that a powerful theoretical structure runs through Aristotle's zoological corpus.[5] And while it may be the first book of the *PA* that states the organizational principles of the project, there can be no doubt that it is the *GA* that exhibits the strictest and most

Translated from the French by Zoli Filotas.

1 Needham 1934: 21, 24. Cf. Louis 1961: XIII.

2 Barthélémy-saint-Hilaire 1887 assigned it to the *PA* (CCLXI). Galen, however, cites it as Book 5 of the *GA* (*On Semen [De semine]* 127, 51–56 De Lacy). See Leunissen and Gotthelf 2010: 325–56.

3 Balme 1992: 127.

4 Van der Eijk 2005: 35.

5 See, for instance, Lennox 2004.

continuous attempt to proceed methodologically, from the beginning to the end of its five books. This is not a matter of the division of the work into chapters, which were added by later editors and are often misleading (especially in *GA* 1),[6] but rather of a set of conceptual distinctions and relationships established by Aristotle himself. There are many such distinctions, they come in several kinds, and some are more important than others. To begin with, Aristotle often arranges his discussions chronologically. Such arrangements play an important role in unifying the treatise, and they are rooted in the very topic of the work, since generation is itself a chronological phenomenon. There are also the contrasts between that which is common to all animals and that which is specific to some of them, between the great zoological kinds, and between different modes of reproduction. Moreover, there is the primacy of blooded over bloodless animals (and of human being over the other blooded animals), and the difference between explanations using efficient and final causes on the one hand, and material and necessary causes on the other. Together these distinctions provide the framework underlying the treatise as a whole, and they organize it by imposing a highly determinate and hierarchical structure.

In the pages to follow, I would like to focus on another distinction: that between the study of parts on the one hand and of generation on the other. This distinction, I will argue, is fundamental to both the structure of the zoological corpus and the development of the *GA*. My thesis is neither historical nor philological. Aristotle might well have gathered together materials from various sources in order to assemble his treatise, and some of his methodological signposts may only have been added at a later stage.[7] But whatever its history, the treatise is clearly organized in service of an overarching plan. I hope to show that this plan reveals itself particularly clearly in the prologue, which I believe may be taken seriously as a set of guidelines indicating the place of the treatise in the zoological corpus as well as its overarching goal and structure.

The Prologue to the *GA*

The prologue to the *GA* (1.1, 715a1–18) is one of the rare texts in which Aristotle sketches the large-scale organization of his corpus. It is

[6] Beullens and Gotthelf 2007: 469–513; Gotthelf 2012: 316n31.

[7] The text includes a number of passages that scholars have rightly or wrongly deemed interpolations, especially in Book 1, but this has no bearing on the treatise's overall structure.

comparable to *Meteorology* 1.1, although of course that text operates at a higher level still. The prologue to the *GA* has often been studied for its lessons about the relationship between the great treatises on animals: the *PA* and the *GA*, the *IA*, and the *HA*. But its bearing on the structure of the *GA* itself has attracted less attention, even though Aristotle explicitly links these different points. Here is a translation of the entire prologue:

> (1) Since it was stated about the other parts in animals, both in common and separately for each kind concerning their distinctive features, in what way each is on account of this sort of cause – I mean this cause: the one for the sake of something – we admit there are in fact four kinds of causes: *that for the sake of which as end* and the *definition of the substantial being* (these should be taken pretty much as one thing) and, third and fourth, the *matter* and *that from which there is the principle of motion*; now, it was stated about the other causes (given that the definition and that for the sake of which as end are the same thing, and given that the matter for animals is their parts: the non-homogeneous parts for every animal as a whole, and the homogeneous parts for the non-homogeneous parts, and what are called elements of the bodies for the homogeneous parts), there remain the parts that contribute to animals their coming to be, about which nothing was determined earlier, and, in regard to the moving cause, what principle it is. (2) But actually, to inquire about this cause and about the coming to be of each animal is in a way the same thing; (3) this is why our argument has assembled a unified whole by putting these parts last in our account of the parts, and, in our account of generation, the principle next after these [parts].[8]

The structure of this preamble is complex, and its style – especially that of the last sentence, our (3) above – is polished.[9] It suggests an elegant and systematic articulation of Aristotle's zoological works, and explains how he has gathered everything not yet discussed into a single text, the *GA*.[10] In this way, I will argue, Aristotle indicates the design of the treatise as a whole.

[8] This translation is adapted from the forthcoming translation by Allan Gotthelf and Andrea Falcon. The numbers in parenthesis are mine. In the original Greek, there is a post-positive δέ at the beginning of the work, as there is in the *GC*. If this is not a so-called "inceptive" δέ, we might wonder where to find the μέν to which it corresponds. The two possibilities would be the final sentences of the *PA* and the *MA*, both of which introduce the *GA*, but neither contains a μέν. See Brunschwig 2004: 26–8. I defend my translation of the final sentence in this chapter.

[9] This is noted by Balme 1992: 127.

[10] At 715a15–16, I translate λόγος as "argument," taking this to include the entire discussion to follow. For this translation see Chapter 1 in this volume. Cf. Bonitz 1870: 433b30–47. (In this context, Aristotle could not have used the term in the plural.)

Aristotle distinguishes between two sorts of objects of study: the parts of animals and their causes. Over the course of the first, fourteen-line sentence (715a1–14), he shifts from the former to the latter – from the parts (715a1: περὶ τῶν ἄλλων μορίων) to the causes (715a7–8: περὶ μὲν οὖν τῶν ἄλλων). The first sentence, labeled (1), makes two conceptual moves: first, it recalls, in a scholarly manner, the doctrine of the four causes (715a3–7); second, it shows how this doctrine directs the internal structure and organization of the zoological treatises (715a8–11). Aristotle begins by stating that he has finished his explanation of all nongenerative animal parts – not only in terms of their final but also their formal causes, since the two coincide in generated beings.[11] This refers at least to the achievements of the *PA*. Aristotle announces that he has also finished his discussion of the material causes of animals, namely their homogeneous and nonhomogeneous parts and the bodily elements.[12] And since he has already dealt with three kinds of cause – final/formal and material – he concludes that all there remains to discuss is the moving cause.[13]

With regard to the parts of animals, Aristotle distinguishes between what he calls the "other parts" and "the parts that contribute to animals their coming to be." The latter, he says, are all that remain to discuss, and in fact he has said nothing about them so far. "Nothing" is an exaggeration, but not an unreasonable one, since when he mentions generative parts in the *HA* and the *PA* it is only to make bare statements, still awaiting explanation. Indeed, in the *HA* and the *PA* Aristotle explicitly postpones the study of the generative parts, whether homogeneous or nonhomogeneous.[14] The two topics that remain to be discussed in the *GA* are therefore the moving cause and the generative parts. It follows that the *GA completes* the study of animals for two reasons: because it treats the *moving cause*, and no other causes remain, and because it bears on the

[11] *Phys.* 2.7, 198a25–26; *Meta.* 8.4, 1044a36–b1.

[12] Aristotle arranges these concepts hierarchically: he reduces homogeneous parts to their function as matter for nonhomogeneous parts. They may also be nutriment or residue, like semen (*PA* 2.2, 647b20–29).

[13] Some translators have detected a redundancy in the question asked at 715a13–14: "περὶ αἰτίας δὲ τῆς κινούσης τίς ἀρχή." Balme 1992: 21, Bos 2003: 148, and Gotthelf 2012: 92 understand Aristotle to be looking for the "source of the moving cause." This translation is problematic. Even in Drossaart Lulofs' text, the sentence means simply that it remains to be shown what is the moving cause, not what is *the principle of the moving cause.*

[14] *PA* 2.3, 650b10; 2.7, 653b16; 2.9, 655b25; 3.5, 668a6; 3.14, 674a20; 4.4, 678a19–20 and a25; 4.10, 689a10, 12 and a20; 11, 692a16–17; 12, 693b25 and 695a27–28; 4.14, 697b29. The same is true of *HA* 3.22, 523a14.

generative parts, and no other parts remain. In this way, the *GA* completes Aristotle's entire zoological project.[15]

Aristotle then adds two sentences that complicate this conclusion. In (2), he explains that he will effectively kill two birds with one stone because studying the moving cause of each animal's generation is "in a way" the same thing as studying generation itself. What is important here is not just that Aristotle qualifies the identity claim (715a15: τρόπον τινὰ),[16] but above all that he now introduces generation, a concept not previously at issue. So Aristotle finds himself with three topics to address in the treatise: (1) the generative parts, (2) the moving cause, and (3) generation. These three are reduced to two through the quasi-identification of the study of generation with the study of its principle. Finally, in sentence (3), Aristotle explains how, by conferring order on the two topics that remain for him, he has constructed a unified whole. I will return to this final sentence in due course.

Aristotle thus confers order and unity on his study by tying together one program based on the parts of animals and another defined by the three causes. His use of these two organizational principles must be understood in light both of the treatises themselves and of the other theoretical strategies he uses to structure his corpus.

PA and *GA*

Some scholars challenge the authenticity of the prologue because of the correspondence – attested nowhere else in the corpus – that it implies between the four causes and the zoological treatises. This correspondence raises several problems. According to Balme, it is simply false that the *PA* appeals only to formal/final causes. The treatise, he writes, discusses other kinds of cause, including the moving cause; moreover, for Aristotle, the physicist *must* use all four causes.[17] This worry cannot be addressed without considering exactly what treatise Aristotle is referring to. There are two possible readings, one broad and the other narrow.

15 A clarification is in order: in studying the generation of animals, as we will see, the *GA* certainly completes one part of the zoological project (*PA* 1.1, 640a10–19, 640a33–b4). But it does not complete the entire biological corpus, as we immediately realize if we consider *Long.* 6, 467b5–10 and the treatise on plants to which Aristotle refers several times over the course of the *GA*. On the existence of the *On Plants*, see Falcon 2015.

16 I say more about this later in this chapter.

17 Balme 1992: 127. At *Phys.* 2.7, Aristotle does write that the physicist must answer the question *why* with regard to all four causes. He does not, however, say that the four kinds of explanation must be given at once or in a single treatise.

The first sentence of the prologue distinguishes between generative parts and other parts. We might understand the latter group to include the parts discussed in the *PA* together with the locomotive parts studied in the *IA*.[18] Both sets are after all explained in terms of final causation.[19] Later in the prologue, Aristotle claims to have discussed not only the formal/final cause but also the material cause, explaining that for animals this consists of the nonhomogeneous and homogeneous parts and "what we call the elements of bodies."[20] Two of the zoological treatises contain material that answers to this description, the *HA* and the *PA*. Thus, Pierre Pellegrin argues that the prologue to the *GA* establishes a long sequence composed of the *HA*, the *PA*, and the *GA*.[21] He claims the prologue shows that in each of the treatises Aristotle privileges one of the causes. In his view, the *HA*, since it studies the parts of animals, is dedicated to the explication of material causation, the *PA* uses final/formal causation, and the *GA* takes up efficient causation. It is true that at the beginning of *PA* 2.1, Aristotle explains that in the *HA* he has just shown for each type of animal how many parts it has and what those parts are (646a9–10). But we might nevertheless wonder how well Aristotle's account in the prologue captures the *HA*, a work that Aristotle does not mention there, though he cites it frequently elsewhere. In the prologue, Aristotle extends material cause to include not only homogeneous and nonhomogeneous parts but also "what we call the elements of bodies." And while the *HA* dwells at length on the nonhomogeneous and homogeneous parts of the great zoological kinds (*HA* 1–4), neither does it limit itself to the parts of animals nor does it ever discuss the elements of bodies. Moreover, it is not clear how it would even be possible for the *HA* and the *PA* to deal entirely separately with the parts of animals (that is, their matter) on the one hand and their final causes (explaining why they have those parts) on the other. And although in the *HA* Aristotle expounds at length on the parts out of which animals are composed and establishes connections between their attributes, he never conceives of them *as the material causes of animals*. They only become material causes when, as in the *PA* and the *GA*, Aristotle makes the fact

[18] Peck 1942, in a note to 715a16 and on page VII.

[19] *IA* 1, 704a4–6. The ending of the *IA* reads like a general conclusion concerning the parts of animals (714b20).

[20] Aristotle often uses such circumlocutions. Cf. Bonitz 1870: 702b2–7.

[21] Pellegrin 1982: 171 ff., and (with different arguments) Pellegrin 1986: 157. Since the *HA* is a repository of premises about correlations that the *PA* and *GA* set out to explain, and since the premises of a syllogism derive from material causes (*Phys.* 2.3, 195a16–20; *Meta.* 5.2, 1013b17–21), Pierre Pellegrin argues that the *HA* provides material causes. But in this sense it does not concern the matter of animals, as does *GA* 1. See also Rashed 2004: 188–9 and 2005: CLXX.

that they belong to certain animals the object of an explanation. And finally, throughout the *GA*, the *HA* is not described as an *earlier* treatise, but rather an auxiliary study, often associated with the *Anatomai*, to which readers may refer for clarification.[22] Thus the *HA* is not so much the first book in a series as a text to be read in parallel. This is confirmed by the fact that, in the zoological works, Aristotle cites the *HA* more often than any other treatise.[23] As Lennox puts it, the *HA* is a "reference work," rather than a "lecture course." It has a similar status to the collection of *Anatomai*.[24]

We might then be tempted by a broader interpretation of the prologue, according to which Aristotle is referring not to one zoological treatise but to several treatises on natural philosophy including the *GC* and *Mete.* 4, both alluded to in the *GA*.[25] The horizons of the prologue, however, are strictly zoological. It is about the causes and the parts *of animals*, and the arrangement of matter into a hierarchy is meant only to lead up to their nonhomogenous parts. Thus, there are several reasons to think that Aristotle has in mind an even smaller and more limited purview: just the *PA* and the *GA*.[26]

When Aristotle says that he has already dealt with the material causes of animals, we may take this as a reference to what he says about matter in the explanatory context of the *PA*.[27] It is well known that the *PA* bears on nonhomogeneous and homogeneous parts, and that *PA* 2 begins with a rather lengthy series of comments on the "composition" of the nonhomogeneous parts out of the homogeneous ones and of the homogeneous parts out of matter (*PA* 2.1–2). Moreover, elsewhere in the treatise, Aristotle provides the material composition of the homogeneous parts he studies. The fibers suspended in blood are made of earth (*PA* 2.4, 650b18), serum is the component made out of water (*PA* 2.4, 651a17), the brain is a

[22] The *PA* treats it in the same way (see *PA* 2.3, 650a31–32). For more on Aristotle's references to the *Anatomai*, I refer the reader to Chapter 13 of this volume.

[23] Eleven times in the *PA* and twelve in the *GA*.

[24] Lennox 1996: 245; Carbone 2011.

[25] *Mete.* 4 is very much in the background of the treatise (compare *GA* 2.2 and *Mete.* 4.1, 379a15–18), and it is the object of many specific references (compare *GA* 2.6, 742b6–7 and *Mete.* 4.4–7; *GA* 5.4, 784b10 and *Mete.* 4.1, 379a15–18). In *GA* 4.3, 768b23–25 Aristotle refers explicitly to *GC* 1.6–7. On the role of the *GC* in relation to the rest of the corpus, see Burnyeat 2004a: 17–18 as well as Rashed 2005: CXL–CLXXVI. The latter underlines the difference between the matter of the *GC* and of *HA–PA* 2.

[26] I set aside the place of the *IA*, which is no doubt a later addition, but which nevertheless belongs to the same sort of inquiry as *PA* 2–4. See Rashed 2004: 189 and Burnyeat 2004b: 21–2 on the notion of an "evolving overall plan."

[27] It is worth noting that in the prologue, Aristotle does not mention the material cause *before* the final cause, but *after*. This is a sign that he is not necessarily taking the treatises in order.

mixture of water and earth (*PA* 2.7, 653a20–21), and so on.[28] Explanation in terms of material causes (in this case the bodily elements) is therefore also present in the *PA*. The prologue, then, is perfectly suited to addressing the *PA* and the *GA* alone: final and material causes in the *PA* and moving causes in the *GA*.[29] This assignment of etiological roles, however, is not without problems of its own, for the prologue suggests that the *PA* studies *all* of the parts except the generative ones, which are reserved for the *GA*. In considering these two points, it will be useful to recall the research method established in *PA* 1.

Although it does not mention the *GA, PA* 1.1 pretty clearly links the two treatises, or at least the types of inquiry. Recall Aristotle's explanation of this point in *PA* 1.1:

> We should also not forget to ask whether it is appropriate to state, as those who studied nature before us did, how each thing has naturally come to be, rather than how it is. For the one differs not a little from the other. It seems we should begin, even with generation, precisely as we said before: first one should get hold of the phenomena concerning each kind, then state their causes. For even with house-building, it is rather that these things happen because the form of the house is such as it is, than the house is such as it is because it comes to be in this way. For generation is for the sake of substantial being, rather than substantial being for the sake of generation.[30]

Aristotle asks whether we should proceed as the ancients did, describing how each thing *comes to be*, rather than the way it *is*. His answer seems rather unresponsive: he writes that "even with generation" we must start by positing *phainomena*, and only then establish their causes. Although this does not quite answer the question he has asked,[31] Aristotle is evidently prescribing a single method, both when we study completely realized beings (i.e., adults) and when we study their generation and growth. Later, when he finally returns to his original question, he writes that we must proceed in a particular order: we must start with a being's *ousia*, then establish that the *ousia* is the cause of certain parts, or that it is impossible for an animal not to possess those parts, or instead show that a given part

[28] *PA* 3.4, 665b6; 3.5, 668b10.

[29] This is also Michael of Ephesus' reading (In *GA* [*CAG* 14.3] 1, 3–5). He writes that the *PA* seeks out the final and material cause of generation, while the *GA* bears on its moving cause.

[30] *PA* 1.1, 640a10–19. The translation of this passage hangs on the phrase "καὶ περὶ γενέσεως" ("even with generation") at 640a15. I use Lennox's translation, which (like Balme 1992 and Pellegrin 2011) reads καὶ as an adverb.

[31] It is closer to the question of 639b5–10, which does not bear on generation.

constitutes in itself a good or an end for the type of animal that possesses it.[32] He then applies what he has just said to the case of generation, as follows:

> And since it is such, its generation necessarily happens in this way and is such as it is. (This is why this part comes to be first, then that one.)[33]

Thus, it is impossible to know the order of the generation of a being's parts before knowing the being itself. Knowledge of *genesis* presupposes knowledge of *ousia* because "*genesis* is for the sake of *ousia*" (640a18), not *vice versa*. Thus, Aristotle combines two kinds of order: (1) an epistemic order according to which we must begin by positing *phainomena* and then going on to establish their causes[34] (if we accept the translation of the text given above, this method *also* applies to generation) and (2) an ontological/gnoseological order, according to which we must first study *ousia* and then use it as the basis for our study of *genesis*. But it is necessary to connect the two orders of inquiry, and that is what Aristotle does. The distinction between establishing the *phainomena* and explaining them applies just as well to *ousia* as to *genesis*, just as well to completed being as to its generation. Following this program, *HA* 1–3 arranges the *phainomena* for all the homogeneous and nonhomogeneous parts, including the generative parts; *HA* 5–7 lays out the *phainomena* relevant to generation according to the kinds of animal;[35] *PA* 2–4 takes up the explanation of these parts in completed beings; and the goal of the *GA* is then to explain the ordered generation of these parts. The *PA* has ontological priority, since it establishes the causes of the parts of a substance – the animal in entelechy – whereas the *GA* seeks out the causes of the generation of that substance.[36] Aristotle famously summarizes all of this with the formula "a human being is generated from a human being." It is because the former – the father – is thus that the latter – the son – is thus engendered. The *PA* explains the former, whereas the *GA* goes on to explain the *genesis* of the latter.

Thus in *PA* 1.1, Aristotle sets up *PA* 2–4 and the *GA* as a neatly ordered, hierarchical dyptich, with *ousia* on one side and *genesis* on the other.[37] Someone might object, however, that in the light of the

[32] On these lines (640a33–b1) – which are disputed – see Lennox 2001c: 134–5.

[33] *PA* 1.1, 640b1–3. Trans. Lennox.

[34] On these two stages of the Aristotelian scientific method, see Lennox 2011.

[35] I will not try to address the authenticity and status of *HA* 10, on sterility. See van der Eijk 2005: 259–75

[36] See Code 1997.

[37] Gotthelf (2012: 39 and note) vigorously expresses this "division of labor."

prologue to the *GA* things appear to be rather different, and that the zoological method established in *PA* 1.1 does not map in any obvious way onto the prologue to the *GA*. On this view, the division of inquiries according to types of causes is unique to the prologue, and the distinction it makes between parts and generative parts is inconsistent with the methodology of *PA* 1.1.

But I believe that the division of etiological labor in *GA* 1.1 is compatible with the principles *PA* 1.1 uses to organize the corpus. It would be most unexpected for Aristotle to contradict himself on this point. Even if Aristotle added *GA* 5 to his study of generation well after the rest of the work,[38] the fact that he refers explicitly to *PA* 1.1 (640a13–19) makes it highly unlikely that the prologue to the treatise includes any real innovation with regard to theoretical arrangement of the corpus:

> As it was stated according to the explanatory principles in the opening arguments, it is not because each thing's coming to be is of a certain sort that on account of that it is of a certain sort, but rather, in as many things as are ordered and definite products of a nature, it is because they *are* such things that they come to be of this sort. The reason is that the coming to be follows upon the substance and is for the sake of the substance, but this does not follow on the coming to be.[39]

PA 1 does indeed appear to be "a philosophical prolegomenon to the entire investigation of animals."[40] And the correlation between treatises and causes is not new; rather, it is a different way of expressing the same distinction between the study of *ousia* and of *genesis*. The formal/final cause is privileged in the study of perfected beings – which is to say in the *PA*, which studies adult animals and does not concern itself with generation. When Aristotle writes in the prologue that the study of generation is "in a way" the same as the study of the moving cause, this is because there can be no explanation of becoming that does not appeal to the cause that sets it in motion. This is not specific to generation – it applies quite generally to all physical explanations.[41] Thus if the *PA* concerns itself with *ousia* and the *GA* with *genesis*, it is only natural that the former privileges the formal/final cause while the latter turns to the moving cause: the cause of *genesis*, which is to say the passage from potentiality to actuality.

[38] If this were true, it would in no way undermine its continuity with the rest of the treatise.

[39] *GA* 5.1, 778b1–6 (trans. Gotthelf and Falcon).

[40] Lennox 2010: 58, 76; Rashed 2004: 191.

[41] *Meta.* 7.17, 1041a31–32 (on the distinction between inquiry into the cause of being and the cause of becoming); 8.6, 1045a30–31, 1045b20–22.

The Problem of the Placement of the Study of the Generative Parts

The real problem here is actually the distinction between parts on the one hand and generative parts on the other. It must be admitted that the two treatises that have come down to us, the *PA* and the *GA*, do not reflect the methodological distinction promised in *PA* 1.1. Indeed, the prologue to *GA* 1 states precisely that Aristotle has *not* respected the prescribed order: *ousia* in the *PA* and *genesis* in the *GA*. He writes that we will not study the generative parts alongside the other parts in the *PA* as we should. The fact that the generative parts are themselves parts is not the only reason why they should have been studied in the *PA*. The defining feature of *PA* 2–4 – a feature that justifies its early placement – is that it explains the parts of completed beings. And the generative parts studied in the *GA*, both homogeneous and nonhomogenous, belong to completed beings. Therefore, they call for the same kind of explanation as the other parts treated in the *PA* and should have been treated there.[42] We might also have imagined them in a treatise of their own instead, like the locomotive parts in the *IA*. Ontologically speaking, in any case, they ought to have been studied with the other parts. But, as Aristotle explains in the prologue, although they are in fact studied in what we might call a "free-floating" section of the *PA* within the *GA*, namely *GA* 1.3–22,[43] they are studied in the *GA* nevertheless.

This placement *can* be explained, but we must not underestimate how problematic it is. Indeed, the prologue is both a sign of the problem and an attempt to solve it. At the beginning of the *GA*, Aristotle finds himself faced with a leftover topic from the *PA*, generative parts, and also with the study of generation itself, which is to say the orderly development of the various parts of a living thing into a complete substance. To better see the difference between these two topics, we may recall the ambiguity of the Greek word γένεσις. Even in the biological works, γένεσις in the singular preserves its general meaning of becoming complete (becoming *x*), insofar as this differs from *ousia*.[44] In the biological context, this phenomenon includes the fact that procreation or engendering occurs,[45] the formation of the embryo, what we call "reproduction" (both sexual and asexual), and birth.[46] To refer specifically to birth, Aristotle modifies

[42] *GA* 2.1, 735a17–19. The defining feature of a complete being is precisely the power of generation.

[43] I defend this partition later in this chapter.

[44] See *Phys.* 2.1, 193b12–13.

[45] This is the most precise meaning of γέννησις.

[46] Bonitz 1870: 149a3. Rashed 2005: LVI ff.

γένεσις with the adverb θύραζε, "outside,"[47] for γένεσις can also mean the generation of the heart, where the nutritive soul is located. Aristotle defines this sort of generation as "the first participation of the nutritive soul in heat."[48] We should therefore understand birth as a "transfer" from one type of generation to another – from intrauterine development (which itself has several stages) to the maturation of an adult organism.[49] In the plural, the word can also refer to modes of reproduction.[50] The *GA* deals with all of these senses of generation: from the creation of seed within adults to the entry of newborns into the world. From among these meanings, it is possible to distinguish two basic orientations: what might be called the "action of generation" on the one hand and the embryogenesis in general on the other (from the first contact between generative movements and menses to the orderly development of parts, to birth).

In an important passage from *PA* 1.5 (645b33–35) that is unfortunately not as clear as we might hope, Aristotle explores the first of these orientations, describing generation and mating as actions (πρᾶξεις).[51] He explains that bodily parts are naturally directed toward the functions and actions for the sake of which they exist. Reproductive organs exist for the sake of actions (mating, sustaining an embryo) while each of the homogeneous parts (semen, menses, milk) has a function. Generation is a function common to all living things;[52] plants and some animals have no function except reproduction;[53] and reproductive organs have a function, just as semen does.[54] Thus, where the explanatory project of *PA* 2–4 is to spell out the function of the homogeneous and nonhomogeneous parts, the functions of reproductive organs, semen, menses, and milk fall under the purview of the *PA*. These parts could have been studied in the *PA*, for the function or action of each is not generative in the sense Aristotle has in mind when he contrasts *genesis* and *ousia:* it is not the formation of a substance (embryogenesis), but rather the activity of an adult substance (mating), or the functioning of the homogeneous parts of an adult substance (semen and menses). Thus, there are indeed reasons that Aristotle might have written a separate treatise on the functions of generative parts in general. If we conform to the program of the *PA*, the only topic falling strictly to the *GA* is *genesis* as embryogenesis, not the functions of the generative parts of adult beings. Thus the *GA* ought to give an account of

[47] *GA* 4.10, 778a10–11 and, *a contrario, GA* 4.9. [48] *Juv.* 2, 479a29–30.
[49] *GA* 4.8, 776b1–2. [50] For example at *HA* 5.1, 539a1.
[51] See also *HA* 5.2, 539b20 and 8.1, 589a2–5. [52] *GA* 1.23, 731a30. [53] *GA* 1.23, 731a24.
[54] Organs: *GA* 1.2, 716a23–27; semen: *GA* 1.18, 724a16.

the *phainomena* of *HA* 5–7 (on generation), but not those of *HA* 1–3 (on animal parts). All of this establishes the importance of the prologue – especially its last lines, which finally explain how Aristotle unites the study of the functions of generative parts with the study of generation itself.

Generative Parts and Generation

Let us turn, then, to the final sentence (3), which allows a particularly wide range of interpretations:

> διόπερ ὁ λόγος εἰς ἓν συνήγαγε, τῶν μὲν περὶ τὰ μόρια τελευταῖα ταῦτα, τῶν δὲ περὶ γενέσεως τὴν ἀρχὴν ἐχομένην τούτων τάξας.

Once again, Aristotle distinguishes between two domains – the parts of animals and generation (i.e., the moving cause) – and indicates how he has taken first "these things" (ταῦτα), and then "the principle" (τὴν ἀρχήν), and arranged them one after the other (ἐχομένην). Translators have generally understood Aristotle to be moving from one level of reference to another, so that "these things" refers to the *parts* that still need to be discussed, while "ἀρχή" designates the beginning of the *discussion* of generation.[55] But since Aristotle claims to have drawn these two things together into one, it would be preferable to find an interpretation according to which the two terms belong to the same level of analysis and refer to entities of the same domain. This is why in their translation Platt and Louis consider "these things" to be *discussions* of the parts, rather than the parts themselves.[56] If we take "these things" to be the "parts," we might instead think that "ἀρχή" refers not to the beginning of the *discussions* of generation, but to the *principle of generation* – to wit, the moving cause.[57] In fact, this last solution must be correct, considering what Aristotle has just said (at 715a11–14): that with regard to the parts, those used for generation remain to be discussed, and that with regard to the

[55] Peck 1942 translates as follows: "our treatise has brought the two together, by placing these parts at the end of our account of the parts, and by putting the beginning of the account of generation immediately after them"; Balme 1992 translates: "our exposition has brought them together, putting these parts at the end of our account of the parts, and the beginning of our account of generation next after them."

[56] Platt 1912 translates: "arranging the discussion of these parts last, and the beginning of the question of generation next to them"; Louis 1961: "Il achève l'examen des parties par celui des organes de la génération et donne à l'étude de la génération le début qui fait suite à celle des autres parties."

[57] I claim that Aristotle could in fact have written "moving cause" here as he does at 715a12. He no doubt chooses "τὴν ἀρχήν" for stylistic effect (opposing it to τελευταῖα, even though the two terms are of course not at the same level from a grammatical point of view).

moving cause, it remains to be established what principle it is.[58] To the extent that the study of generation and the study of this principle "in a way" come to the same thing, Aristotle declares that he has assembled the generative parts and the principle of generation into a unified whole by arranging them in order. In addition, Aristotle's construction, using two partitive genitives, signifies that just as not all parts are generative parts, generation cannot be reduced to its principle. The study of the principle and of generation are only identical "in a way," since the material cause – the menses – must also be taken into account.[59]

In his commentary, Michael of Ephesus explains that there is a "continuity" between the *PA* and the *GA*, and that the five books of the *GA* are placed after those of the *PA* and therefore ranked below them. In other words, he interprets this sentence as if the "unified whole" in question consisted of the two zoological treatises.[60] But the text suggests that the unified whole is rather the *GA* itself. The *GA* follows the *PA* because the *GA* completes the investigation of the *PA* by studying the generative parts. But this study is for Aristotle also an integral part of the *GA*. And the sequel to this first part of the treatise, he says, concerns generation itself and more precisely the moving cause. Thus, the *GA* is a unified whole assembled from two sections: the study of generative parts and the study of generation, which is also the study of the principle. How is this design realized in the treatise? There are two possible answers, depending on whether we take this preamble to describe just the organization of *GA* 1 or rather the structure of the treatise as a whole. Balme and Gotthelf think that the two studies correspond to the two main sections in *GA* 1: Chapters 3 to 16 study the reproductive parts in the manner of the *PA*, while Chapters 17 to 22 initiate the study of generation.[61]

But, *contra* Balme and Gotthelf, it is hard to see why the "parts" studied last would only be the reproductive organs, or why the "beginning" of the study of generation would consist of Chapters 17 to 21. Aristotle does indeed ask how to identify the principles of generation, and offers an answer at the start of *GA* 1.2. One of the main arguments for Balme and Gotthelf's reading is the reference at 715a12 to the "parts that contribute

58 This translation is permitted by Drossaart Lulofs' text, which reads Z's ἀρχή at 715a14 rather than the αὕτη of all of the other manuscripts.

59 We cannot rule out the possibility that this qualification also opens the door for the explanatory style of *GA* V. See also Gotthelf 2012: 92.

60 Michael of Ephesus, *In GA* [= *CAG* 14.3] 3, 5–10.

61 Balme 1992: 127; Gotthelf 2012: 92. Gotthelf describes *GA* 1.2–16 as a "*PA*-like, three-cause study of the generative parts" (2012: 91). See also Chapter 1 in this volume.

to the generation of animals" (τὰ πρὸς τὴν γένεσιν συντελοῦντα). This phrase, which reappears in *GA* 1,[62] refers to the reproductive *organs*, whether internal or external.[63] This suggests that Aristotle is referring to the discussions in *GA* 1 that address these *organs*, namely Chapters 3 to 16. The parts that contribute to generation, however, are not exhausted by those that Aristotle sometimes chooses to refer to precisely as the organs used for mating.[64] The homogeneous parts, such as semen, menses, and milk, also contribute to generation. For Aristotle, the unique place of semen and menses among the homogeneous parts[65] in no way detracts from their status *as* homogeneous parts.[66] And at the end of *GA* 1.16, Aristotle states that he has just examined the *organs* that animals use for generation, which he had not previously discussed. Now, he writes, he will dedicate himself to the homogeneous parts that remain (and which he has also not yet discussed): semen and milk.[67] It seems overly restrictive in this context to think that the expression in the prologue merely designates the male and female reproductive *organs*, excluding the homogeneous parts (semen, menses, and milk).

Moreover, if we adopt the reading according to which the plan from the prologue is realized only in *GA* 1, we find ourselves with two further problems. (1) Since the prologue ties together the *PA* and the study of generation in its entirety, it is hard to see why the plan of the final sentence would refer only to the first book of the *GA*. (2) Since Aristotle announces that he aims to bring his analysis to bear on the parts he has not discussed earlier, why would he restrict the reproductive parts just to the *nonhomogeneous* parts, given that what remains after the *PA* are both reproductive organs and homogeneous parts like semen and menses? In the *PA*, his forward references to the *GA* concern the organs no less than semen, milk, or menses.[68] It is for this reason that Michael of Ephesus takes the study of the generative parts to include the examination of both nonhomogeneous and homogeneous parts – and I think he is right.[69]

I believe that it is possible to defend another reading of this final sentence. Aristotle's plan does not concern just *GA* 1 but the whole of

62 *GA* 1.14, 720b1, 3.
63 *GA* 1.16, 721a26. Along the same lines, see *HA* 3.1, 509a29 and *PA* 4.4, 678a22.
64 *GA* 4.1, 764a30–32 and 1.15, 720b35.
65 *PA* 2.2, 647b20–29. On this point, see Chapter 6 in this volume.
66 *HA* 1.1, 487a1–10; *PA* 2.9, 655b 23–27 and 4.10, 689a 8–17; *GA* 1.18, 724b 29.
67 *GA* 1.16, 721a26–30.
68 Semen: *PA* 2.7, 653b13–18, 2.1, 655b23–27; reproductive organs: *PA* 4.4, 678a22, 4.10, 689a17–20, 4.12, 695a27.
69 Michael of Ephesus, *In GA* [= *CAG* 14.3] 1, 16–20.

the treatise. But, crucially, Aristotle says that he has made the two sections a *unified whole*: although one follows the other (ἐχομένην), *they are not juxtaposed*.[70] The *GA* does not consist of an application of the method of the *PA* to the generative parts (*GA* 1), and then of a study of the principle of generation. *GA* 1 is not Book 5 of the *PA*; rather, it is an integral part of the study of generation. It is this integration of *GA* 1 into a study of generation that leads Aristotle to study the generative parts in the context of the *GA*, rather than in the explanatory context of the *PA*. If Aristotle specifies that he has constructed a unified whole it is because, as we have seen, the methodology formulated in *PA* 1.1 separates the study of the generative parts from the study of generation for ontological reasons. Their reunification is only possible because, although the study of the generative parts is carried out in light of the final cause (just like all the other kinds of parts in the *PA*), it advances knowledge of the principle of generation. It is worth noting that the study of generation does not begin *after* the discussion of the generative parts, but that, to a remarkable extent, it serves as a guiding thread for the study of those parts, including the reproductive organs no less than the homogeneous parts.

The Integration of the Study of the Generative Parts into the Treatise on Generation

The structure of *GA* 1 shows to what extent the book is not merely an application of the explanatory method of the *PA* to generative parts, but rather results from its integration into the study of generation, understood at this stage not as embryogenesis but as a specific action performed by *completed* animals. In line with the prologue, Aristotle's study of the principle of generation as embryogenesis does not begin until *GA* 2. As for *GA* 1, it is systematically arranged into three nested stages in chiasmus that integrate the study of the functions of generative parts into a study of the principle of generation.

(A = GA *1.1, 715a18–716a2)*

After the preamble, Aristotle begins by linking animal generation with the coupling of male and female. In all species with male and female members, animal reproduction depends on males and females mating. These species

[70] The expression at 715a15–16 is also used at *Pol.* 3.9, 1280b14, and in biological contexts to express how nature "has collected" several functions in one organ (*PA* 3.1, 662a23).

include not only blooded animals, but more generally animals capable of locomotion (with the exception of insects, some of which are not differentiated into males and females). The case of hard-shelled animals (ὀστρακόδερμα), which do not move around and are not distinguished from each other sexually, confirms the rule. The generation of living things is thus sexual in the case of animals, except for certain insects and for shellfish. Among living things, this distinguishes them from plants. Here Aristotle simply asserts the relationship between animal generation and sexual generation, although he will offer a teleological explanation of the phenomenon at the end of the book (1.23).

(B = GA *1.2, 716a2–17)*

It follows that male and female are "principles of generation." Aristotle offers up what is for now a hypothesis explaining how each of them satisfies this description: the male is "that which possesses the principle of movement and of coming to be"; the female is that which has the principle of the matter (716a4–7).[71] Since male and female are principles of generation because of the secretions they emit (either within themselves or externally), demonstrating this hypothesis requires knowledge of the formation and function of the seed (σπέρμα) in males and females.

(C = GA *1.2, 716a17–b12)*

The difference between male and female is not restricted to function (the male emits an external secretion, while that of the female is internal; the male function is therefore to engender within another, the female within herself). That functional difference is inseparable from an anatomical difference, since different powers presuppose different organs and therefore different parts. Those parts are the organs responsible for the mating and birth of male and female animals.

The following chapters in *GA* 1 (3–23) answer to these three parts in a remarkably strict chiasmus: A B C – C′ B′ A′. Aristotle successively considers the male and female reproductive organs (C′ = *GA* 1.3–16), and the nature and formation of male and female secretions, which allows

[71] When he specifies the ways in which male and female are principles of generation, namely as moving cause and as matter, Aristotle offers a hypothesis that is relevant beyond the *sexual* generation of animals. Spontaneous generation (treated at *GA* 3.11) also appeals to a material and efficient principle.

him to demonstrate the claim initially advanced as a hypothesis (B′ = *GA* 1.17–22), and finally the reason that the male and female principles are not separated in plants (A′ = *GA* 1.23). It is clear that the study of the two kinds of generative parts serves Aristotle's research on the principles of generation. In other words, just as he announces in the prologue, Aristotle has constructed a unified whole: by starting the study of the principle of generation with the examination of generative parts, he integrates those parts into the heart of his study of generation. All the way through *GA* 1, his question is how the generative parts contribute to generation: How do the reproductive organs (and their location and orientation) contribute to generation? How do homogeneous parts contribute to generation? In both cases, answering the question amounts to identifying a function. Thus, *GA* 1.3–22 adopts the same style of explanation as *PA* 2–4, the search for a final cause. But its inquiry is subordinated to the general project of the *GA*, disclosing the moving cause of generation.

(C′ = GA *1.3–16)*

This section of *GA* 1 studies the organs corresponding to the male and female powers, either the means by which animals mate or organs that otherwise contribute to generation. It also contributes directly to Aristotle's inquiry into the principle of generation. First of all, (a) it allows him to eliminate other candidates for the status of principle, such as testicles. Not only do some animals lack those altogether, he argues, but they are not linked with the spermatic ducts and they must therefore be unnecessary for generation.[72] Aristotle's description of the double fold of the spermatic duct gives him an opportunity to invoke the transformation of blood into semen. He indicates that while one part of the duct contains blood, there is already semen in another.[73] (b) This section also introduces the importance of heat and concoction. Semen is concocted during coitus (for live-bearing land-dwellers) or earlier (for fish and snakes); the fertility of the semen depends on its heat,[74] which also explains why it is protected in animals with external spermatic ducts.[75] The position of the uterus must also be explained in terms of its heat.[76] (c) Finally, in accounting for the variety of zoological groups, Aristotle rapidly dissociates the semen itself from the generative principle. Aristotle inquiries into the nature of the emission in the case of mollusks and crustaceans – is it semen, some other part, or a power?[77] Semen as such is not the principle of generation. Insects exemplify the case of generation without

[72] *GA* 1.4, 717a17. [73] *GA* 1.6, 718a11. [74] *GA* 1.7, 718a20–27. [75] *GA* 1.12, 719b1–2. [76] *GA* 1.8, 718b18–27. [77] *GA* 1.15, 720b30–31.

spermatic emission by the male into the female, since the female introduces one of her parts into the warmest part of the male's body.[78] Thus, while this section of *GA* 1 might seem like little more than an anatomical study, the explanation of reproductive organs in fact contributes to the search for the moving cause of generation.

(B′ = GA *1.17–22)*

The explicit goal of the analysis beginning at *GA* 1.17 is to define the "nature" and the role of semen and menses. There is thus no need to point out that this teleological step (concerning the function of semen and menses) belongs to Aristotle's research into the principle of generation. What is important here is rather the specific advance that this discussion makes in the search for the principle of generation. The fact that not all coupling animals emit semen, and that females have an emission of their own leads to the question of the specific contribution of semen and menses to generation.[79]

(A′ = GA *1.23)*

The last chapter of *GA* 1 looks past the two other sections, back to the earliest analyses in the book. Aristotle now offers an explanation of the distinction drawn in *GA* 1.1 between sexual and nonsexual generation. He does not provide a teleological explanation of the separation of male and female principles, like *GA* 2.1 (731b18–732a11); what he does, more precisely, is explain, in terms of final causation, the *non*separation of the sexes in the case of plants and the necessity for mating in the case of animals. The explanation begins by comparing the case of animals with that of plants: male and female are principles of generation in plants and animals alike, but in animals they are separated and must be brought together for a certain amount of time for the sake of reproduction, while in plants they are united by nature. Why is it, Aristotle asks, that although reproduction requires these powers to be conjoined, as indeed they are in plants, they are separate in animals? In other words, why does nature separate male from female in animals, but not in plants? The answer calls for a teleological explanation: "nature has acted as a reasonable artisan" (713a24). The male and female principles are conjoined in plants because the very essence of a plant is to engender a seed – its only end is reproduction, and there is thus no need for male and female to be

[78] *GA* 1.16, 721a12–13. [79] On the construction of this stretch of text see Lefebvre 2016.

separated, so it is best that they be together. But it is best for animals to mate because this frees them to be more than just embryo generators. They participate in a minimal form of understanding, sensation. Thus, there is no need for a natural unification of the two natural powers; it is only to reproduce that male and female come together, becoming like a plant.

The structure of *GA* 1 clearly shows how Aristotle places his study of the generative parts before his study of the principle of generation, and thereby assembles them together into a "unified whole." But even in the "*PA*-like" inquiry of *GA* 1, Aristotle is already searching for the principle of generation, i.e., its moving cause. So, *pace* Gotthelf,[80] *GA* 1 is not a "*PA*-like inquiry," not even in its first section, devoted to the study of the reproductive parts. Nevertheless *GA* 1 is only the first step in the search for the principles of generation, not because its results are unsatisfactory but because Aristotle has not yet found his way to generation in the sense in which it is the object of the treatise: as the ordered formation of the parts of a living thing, or embryogenesis. The true study of embryogenesis, in which the principle of generation explicitly becomes the object of study, begins in *GA* 2 and is completed in *GA* 3. *GA* 2 therefore marks a fresh start. Without attempting to give a full treatment of the matter, it is possible to make a few remarks. In *GA* 4.1, Aristotle closes his inquiry into generation by declaring that he has simultaneously studied that which is common to all animals and that which is particular to each group.[81] As a whole, *GA* 4 concerns teratology (4.4), pathology (superfetation, the so-called *mola uteri*), and phenomena that arise after the formation of organs specific to the animal and even after its birth (lactation). In this respect, *GA* 4 follows the genetic or chronological ordering that runs through the treatise. But the crucial point is that it begins by suggesting an explanation of the formation of the sexes by means of efficient and material causation (*GA* 4.1). Aristotle effectively comes full circle, back to the beginning of the treatise: he explains the *generation* of the male and female generative parts whose function was studied in *GA* 1 in the style of the *PA*. We may therefore say that by the end of *GA* 4.1 the generative parts have received a complete explanatory treatment – in terms of functions and by means of final causation in *GA* 1.3–23, and then by means of the efficient and material causation in *GA* 4.1. The heart of Aristotle's study of the principle of generation in *GA* 2–3 is in *GA* 2.1, where he raises and resolves the *aporia* that is the very object of the treatise on generation: "starting from seed, how is a plant or an animal engendered, of whatever kind it may be?"

[80] As well as Chapter 1 in this volume.

[81] *GA* 4.1, 765b6–8.

(733b23–24). It is there that Aristotle passes from the exterior to the interior, from the father to the embryo. He shows how there can be continuity between the father, the movements in his semen as they act on the menses, and the formation of the heart, the seat of nutrition that stands at the origin of organogenesis and embryogenesis. At the end of the chapter, he writes, "in response to the difficulty discussed earlier, we have now given the cause that stands as the principle of generation in every case, that which moves first, and which fashions" (*GA* 2.1, 735a27–29).

In conclusion, the prologue teaches us that the *GA* is composed of two main sections: *On the Generative Parts* and *On Generation*. This distinction, which reflects the structure of zoological inquiry, takes nothing away from the profound unity of the treatise: the prologue says precisely that from these two inquiries Aristotle has assembled "a unified whole."[82]

[82] I would like to thank Michel Crubellier and André Laks for their comments on an earlier version of this piece.

CHAPTER 3

Order and Method in Aristotle's Generation of Animals *2*

Mariska Leunissen

Introduction

Generation, Aristotle tells us early in *GA* 2, is something that has been rendered "well and orderly" by nature (*GA* 2.1, 732b32–33: ὡς εὖ καὶ ἐφεξῆς τὴν γένεσιν ἀποδίδωσιν ἡ φύσις). This order in generation is teleological: just like all the natural beings themselves exhibit teleological order and hierarchy,[1] so does the coming to be of their parts.[2] Given the nature and function of each of the parts, suitable matter has to be present first and has to be organized in the appropriate way, and given the teleological and hierarchical connections between the parts of animals,[3]

The present essay has its origin in a presentation I gave in 2010, at a workshop on the methodology of Aristotle's *GA* organized by Allan Gotthelf and Devin Henry. Allan always thought of the *GA* as "one long argument" (thereby borrowing Darwin's characterization of his *Origin of Species*) and he encouraged me to research the tight connections between order and method that run through the second book of the *GA*. I dedicate this paper to Allan's memory. I am also grateful to report that, thanks to Andrea Falcon's efforts and dedication to this project, Allan's ideas about the unity and structure of the *GA* as a whole are published posthumously in this volume as Chapter 1. The most important difference between our essays is that Gotthelf and Falcon argue for a logical, unified narrative ordering of the *GA*, according to which each section is presented when and where it is in the structure of the *GA* with a view to the one long argument that treatise is defending, whereas my focus is more on exhibiting the extent to which the narrative of the *GA* tracks the chronological order of embryological development itself and what principles Aristotle uses when he has to deviate from this order.

1 *Meta.* 12.10, 1075a11–25: "All things are ordered together in some way, but not in the same way – even swimming animals and flying animals and plants. And it [i.e. the order] is not such that there is no relation between one thing and another – there is one: for all things are ordered together toward one (πρὸς μὲν γὰρ ἓν ἅπαντα συντέτακται), but it is as in a household."

2 *PA* 2.1, 640a35–b4: "If one cannot [say this], [one should say] the nearest thing, namely that it is thus either in general (because it cannot be otherwise), or that it is in a good way thus. And these things follow. And since it is such, its generation necessarily happens in this way and is such as it is. This is why this one of the parts comes to be first, then that one (διὸ γίνεται πρῶτον τῶν μορίων τόδε, εἶτα τόδε). And similarly in this way with regard to all things that are constituted by nature."

3 *PA* 1.1, 639b22–30: "The [necessity] that is unqualified belongs to the eternal things, and the one that is conditional also belongs to all things that come to be, as well as to things produced, such as to a house and to any other such thing. For it is necessary that such material is present, if there is to be a house or any other end: and it is necessary that first this comes to be and is changed, and next that,

some parts will need to come be before other parts. In this essay, I discuss the ways in which Aristotle tracks this "order in generation" and other natural hierarchies in the order of his exposition in the *GA*, and in particular in *GA* 2, and show how this concern with following the right order in exposition gets interlaced with concerns about methods for the discovery of these natural orders. In other words, as I hope to show, for Aristotle giving an explanatory account of embryogenesis involves following the natural order of things in that account.

GA 2 and the "Long Argument" of *Generation of Animals* as a Whole

At the outset of *GA*, Aristotle tells us that he has already discussed – presumably in the treatise we know as the *PA* – the formal, final, and material causes of all the parts of animals, except for the parts concerned with reproduction. So, what remains then, is to discuss the generative parts and the efficient causes of the parts of animals, which turns out to be one and the same inquiry (*GA* 1.1, 715a1–15). The discussion that follows in five books tracks roughly the chronological order of embryological development and animal generation and can be divided into four different strands of investigation, each dealing with a subsequent phase of the generation of animals.

The first strand of investigation, comprising *GA* 1.1–2.3, deals roughly with the identification of the first principles and parts of reproduction. The principles of reproduction form, of course, the absolute causal and chronological starting points of the process of animal generation. Since reproduction requires the meeting of these principles, Aristotle also needs to discuss the sexual organs and homogenous parts such as semen and menses involved in reproduction, the separation of the sexes that is found among many animals, and the different modes of reproduction found among different kinds of animals. The second strand of investigation deals with the process of embryogenesis properly speaking from the moment of conception until the *completion of the embryo*, and covers chapters 2.4–3.11. *GA* 2.4–2.8 discuss embryogenesis in the most perfect of blooded animals (I will return to this later), while *GA* 3 covers embryogenesis for each of the remaining kinds of animals. The third strand of investigation, which is collected in our *GA* 4, deals with the

and so step by step up to the end and that for the sake of which each thing comes to be and is. It is the same way too for things that come to be by nature."

explanation of sexual differentiation and heredity and phenomena *surrounding birth*, such as the birth of monstrosities, superfetation, milk-production (a topic that was postponed for later as it was not the right time yet to discuss it in *GA* 1.17–20, which dealt with the homogeneous parts involved with the beginning or reproduction; see *GA* 1.16, 721a27–30), and the position of the offspring during birth. Finally, the fourth strand of investigation, recorded in *GA* 5, deals with affections of the more and the less that differentiate animals and that come to be due to changes during the development of animals that happen immediately *after its birth*, or in later life, or in old age, and that are the most visible in human beings.[4]

This overview is necessarily perfunctory, but I hope it illustrates sufficiently that at the macro-level Aristotle has organized his treatise "chronologically," following the natural order of animal generation and especially of the work done by its motive principles.[5] *GA* 2 completes the first investigation and then includes the most important part of the second one, but the discussions of the different modes of reproduction, of how things come to be from seed, and of how the animal acquires its soul in *GA* 2.1–3 are all methodologically prior to the discussion of embryogenesis in the most perfect, blooded, live-bearing animals that comprises the rest of the chapter. And, as I will show below, the results of this investigation in *GA* 2.1–3 also lay the foundation for what ends up being the ordering principle for the exposition recorded in *GA* 2.4–3.11.

Aristotle's Obsession with the Discovery of Natural Order in *GA* 2.1–3

At the end of *GA* 1 Aristotle has identified the principles of reproduction, discussed the homogenous and heterogeneous parts involved in reproduction (with the exception of milk, which is postponed until later), and has given one explanation – a teleological one – for the separation of the sexes. At the beginning of *GA* 2, Aristotle starts by giving a second explanation of the separation of the sexes (this time one that focuses on the material and

[4] See Leunissen and Gotthelf 2010, for a detailed interpretation of the purpose of *GA* 5 and for a defense of its belonging to the investigation of the *GA* as a whole.

[5] Cf. *GA* 5.1, 778b19–20: "Having made these distinctions, we must speak of what comes next in order." Note that the topic that is discussed immediately after the discussions of phenomena surrounding birth in book 4 is sleep in newborns – a topic that may seem odd at the beginning of book 5, but that is exactly what one should expect when following a rough chronological ordering of topics concerning the generation of animals.

efficient causes; see *GA* 2.1, 731b19–732a25), and it seems that with the principles of reproduction and their function and definition firmly in place, he would now be ready to discuss the process of embryogenesis itself. As it turns out, though, there is at least one problem and three major puzzles that need to be solved before Aristotle can turn to presenting the results of this second strand of investigation. The problem and two of these puzzles[6] reveal an obsession with the identification of natural order, so let me discuss them in turn.

The problem Aristotle raises concerns the existence of different modes of reproduction (*GA* 2.1, 732a25–732b14). One reason why Aristotle cannot delve immediately into a description of the process of embryogenesis is that there appear to be *differences* between the generative processes of different kinds of animals. According to Aristotle, this becomes clear when one looks at the natural variation among the resulting offspring that are produced by different kinds of animals, for different kinds of animals produce offspring that have reached different degrees of perfection at the moment of birth. There are animals that are live-bearing and give birth to offspring that is perfected to the point that it resembles its parents, but there are also animals who produce something undeveloped such as egg-layers and grub-producers. And then even among these categories there are differences: some live-bearing animals produce live young, some lay eggs internally and are only later live-bearing externally, and some animals produce perfect eggs, some imperfect ones. Finally, there are also animals that are generated spontaneously.

Although Aristotle does not explicitly formulate the matter as a puzzle (ἀπορία) or as a problem (πρόβλημα), this differentiation is something that requires explanation, and the next section reflects his search for its cause. "These kinds admit of much cross-division" (πολλὴ ἐπάλλαξις), he says, but none of the divisions Aristotle mentions yield the relevant difference, and thereby the cause: having two versus having four feet, having feet versus not having feet, and neither any of the parts for locomotion themselves are "the cause of this difference" (αἴτιον τῆς διαφορᾶς ταύτης). In fact, it appears that the relevant cause that is responsible for this differentiation is not a formal one. Instead, Aristotle refers to the different levels of perfection of different kinds of animals, and subsequently cashes out this language of perfection in terms of the levels of vital heat and the variations between the material natures of those animals.

[6] The third puzzle about the material nature of semen, discussed in *GA* 2.2, does not exhibit this concern with order.

Table 1 *Aristotle's scale of perfection as based on differences in material natures in* GA *2.1.*

	Animal species	Type of material nature
	Live-bearing (blooded)	***Optimal: hot, moist, and pure***
1a	Internal: humans, mammals, dolphins	Hottest, moistest, almost no earth
1b	Internal eggs: selachia	Hot, moist, very little earth
	Egg-laying (blooded)	***Lacks moisture***
2a	Perfect eggs: birds, snakes, reptiles	cold, dry, earthy
2b	Imperfect eggs: fish, crustaceans, cephalopods	colder, dryer, earthier
	Grub-producing (bloodless)	***Lacks moisture and heat***
3	Insects	Coldest, driest, uses external earth

The more perfect the animal, the more perfect its product of generation is (see especially *GA* 2.1, 733a1–3), and more perfect according to Aristotle means hotter, less earthy (i.e., purer), and moister in its material nature (see especially *GA* 2.1, 732b31–32).

Aristotle distinguishes between three types of material nature (see also Table 1):[7] first, blooded animals that are moist and that have little or no earth; second, blooded animals that are dry and earthy; and third, bloodless animals that are dry and have no earth in their constitution. Within the first two types, differences in the internal heat of animals causes further differentiations in their modes of reproduction: all moist animals are live-bearing (*GA* 2.1, 733a11: "for moisture is ζωτικόν – life-giving"), but the hotter animals among them, such as mammals, are internally live-bearing, whereas the colder ones, such as selachians, first produce internal eggs and are only externally live-bearing afterwards. All dry animals produce eggs, but the hotter animals among them, such as birds, snakes, and scaly reptiles, lay perfect (i.e., hard-shelled, fixed sized) eggs, whereas the colder animals, such as fish, crustaceans, and cephalopods, lay imperfect eggs. The third bloodless group, consisting predominantly of insects, divides further into those that are produced by sexual reproduction and those that generate spontaneously, but Aristotle does not explain the causes of this differentiation here (on this, see *GA* 3.9–11).

Interestingly for our current purposes, Aristotle points out that these "subgroupings" of animals that are differentiated based on their species-specific material natures give rise to a *scala naturae* (*GA* 2.1, 733a32–33:

[7] This paragraph is reprinted, with slight modifications, from my earlier discussion of Aristotle's explanation of the different modes of reproduction in Leunissen 2012: 514–6.

nature organized them well and orderly). In other words, nature has produced a natural hierarchy among these animals, according to which the ones that are hottest, moistest, and not earthy (Aristotle characterizes these animals as "participating in the purer principle"; see *GA* 2.1, 732b29) are *the most perfect,*[8] and the ones that are cold and dry are *the least perfect.* And while the animals in group 1 only take "one process" to produce perfect offspring (that is, offspring that is sexually differentiated and that is complete in terms of its vital and essential bodily parts; the offspring will still grow and acquire parts such as hair and teeth), the animals in group two take "two processes" (i.e., the formation of the egg and then the formation of the perfect offspring coming out of the egg), and the animals in group three do not "reach their end of coming to be" until in their "third transformation" (733b15–16: ἐν τῇ τρίτῃ μεταβολῇ) (i.e., after the grub is formed, then something egg-like comes to be, and then the perfected animal comes to be).

Aristotle emphasizes this point about the hierarchical order of nature as an introduction to his full explanation of how the five different levels of perfection result in five different kinds of offspring, but the statement ends up having normative value as well for the way in which Aristotle ends up organizing his discussion of the details of the process of embryogenesis once he gets to that in *GA* 2.4. In that chapter, Aristotle starts with a brief summary of the causes of the different modes of reproduction mentioned above while adding a description of what counts for a "complete" embryo in each of the three broad groups of animals, and then introduces the discussion of embryogenesis properly speaking as follows:

> Now we must start first from the first [animals]: the first are the perfect animals (τὰ τέλεια ζῷα), and these are the live-bearing animals and first among them the human being (καὶ τούτων ἄνθρωπος πρῶτον). (*GA* 2.4, 737b25–27)

Humans – because they possess the most perfect material nature – exhibit the most perfect mode of reproduction (i.e., their offspring is both sexually differentiated and physically complete in terms of possessing all their vital and essential parts at the time of birth). And for this reason human embryogenesis should be discussed first. For Aristotle, the order of nature has thus normative force over the order of exposition.

[8] Cf. *Resp.* 13, 477a15–25, where Aristotle calls animals with more heat "more valuable" (τιμιώτερα), and ascribes to them a "more valuable soul" and a "more valuable [material] nature," and states that humans are the hottest and purest.

Note that this, of course, is not the only place where Aristotle uses humans or the human body as a guideline or "model" for the order of exposition. For instance, in his introduction to the discussion of the parts of animals in *HA* 1.6, 491a14–26, Aristotle explains that we should start by enumerating the parts on the *human* body, as they are epistemologically speaking "closer to us" and visible to everyone, but also because proceeding in this way will somehow preserve "the proper order" (491a24–25: ὅμως δ' ἕνεκεν τοῦ μὴ παραλιπεῖν τε τὸ ἐφεξῆς).[9] Something similar happens also in the *PA*, where Aristotle begins his treatment of the external, nonuniform parts in *PA* 2.10, 655b28–29 by stating that "we must now speak as if we are again at a beginning, starting first from the primary things," where these "first among firsts" again turn out to be the parts of humans because only their "natural parts are in accordance with nature" (*PA* 2.10, 656a9–13).[10] In sum, Aristotle believes that by taking humans as a roadmap for the discussion of biological phenomena – whether they pertain to parts or processes – he is in fact tracking "the order of nature," because humans are – at least in the sublunary world – the most "natural" and perfect.

Returning then to the *GA*, Aristotle starts his exposition of embryogenesis properly speaking in *GA* 2.4–6 by discussing embryogenesis in the live-bearing animals while putting human embryogenesis at the forefront (*GA* 2.7 discusses the growth of embryos, while *GA* 2.8 is in a sense a digression offering an explanation for a puzzle about sterility mentioned in 2.7). Next, following the natural hierarchy among animals from top to bottom and speaking "in accordance with the path laid down,"[11] Aristotle turns to a discussion of embryogenesis in egg-layers (*GA* 3.1, 749a10–13), first in birds, which are the more perfect egg-layers (*GA* 3.1–2), next in fish, which are the less perfect egg-layers (*GA* 3.3–8).[12] He concludes with embryogenesis in the least perfect animals, the insects (*GA* 3.9–10), and testacea (*GA* 3.11).

[9] See *HA* 1.8, 491a24–26; *HA* 1.15, 494a20–24; *HA* 1.16, 494b19–20; and *HA* 1.15, 494a26–b1; a26–31: "Humankind has more than other animals the upper and the lower [parts] distinguished with a view to the natural locations (πρὸς τοὺς κατὰ φύσιν τόπους): for what is up and below [in humans] has been ordered (τέτακται) with a view to what is the up and the below of the universe. And in the same way [humankind] also has the front and back [parts] and the right and left [parts] in accordance to nature (κατὰ φύσιν)." Cf. *HA* 5.1, 539a10 and 5.2, 539b15.

[10] See *PA* 1.4, 644b20–21; *PA* 3.7, 670b29–30; and *PA* 4.14, 697b28–30.

[11] See *GA* 3.9, 758a28–29: "It remains to speak about [. . .] in accordance with the path laid down (λεκτέον κατὰ τὴν ὑφηγημένην μέθοδον)." Cf. *GA* 1.2, 716a2–3 and *GA* 2.4, 740b11–12.

[12] Note that *GA* 3.6 is another digression that takes the discussion back to the refutation of wrong ideas about bird-reproduction that are similar to the wrong views that had been brought up about fish in the previous chapter.

While the macrostructure of the whole of the *GA* tracks roughly speaking a very general account of embryogenesis, going from first principles to *differentiae* that occur post birth, the exposition in *GA* 2.4–3.11 tracks the order of perfection in nature, by discussing embryogenesis in the most perfect animals first, and in the least perfect animals last.[13]

The second puzzle is in fact explicitly formulated as a "pretty big puzzle" by Aristotle and concerns the question of how living beings come to be from seed (*GA* 2.1, 733b23–735a27). The still very general answer is settled relatively quickly – it must be from something that is already present in the seed and that something must have soul from the very beginning – and it is here that Aristotle returns to the question of order. For how, he asks, does this as-of-yet unidentified part make the other parts? There are two options:

> For either all parts – such as the heart lung liver eye and each of the other ones – come to be at the same time, or successively as in the famous words of Orpheus: for there he says that an animal comes to be in the same way as the knitting of a net. (*GA* 2.1, 734a16–20)

Aristotle rules out the option that parts come to be at the same time by a simple appeal to observation (*GA* 2.1, 734a19–20: ὅτι μὲν οὖν οὐχ ἅμα καὶ τῇ αἰσθήσει ἐστὶ φανερόν): some parts are already present while others are not, and that is not because the latter parts are simply too small to be seen. As Aristotle points out, the lungs are bigger than the heart even though their development starts later and they both develop during the first phase of embryogenesis. So since some parts are earlier, some later, Aristotle needs to settle next whether the earlier parts produce the later parts (i.e., whether the connection is causal) or whether some simply come into being after the other (i.e., whether the development follows a chronological order without the earlier part being responsible for the coming to be of the next).

The first option is rejected immediately on the grounds that it does not fit with Aristotle's thesis that what comes to be in potentiality can only come to be through what already is in actuality, and it is clearly absurd to suggest that the form of the liver already exists in the heart, etc. Aristotle corroborates and works out the details of the second option, according to which parts come to be one after the other but not because of the

[13] Contrast with this Aristotle's discussion of the different modes of reproduction in *HA* 5.1, 539a4–15, where – for pragmatic reasons – he decides to follow the *exact opposite order of perfection*, moving from testacea to insects, then to fish, birds, and live-bearing animals, and lastly to humankind, which involves the longest discussion (but see again the use of ἐφεξῆς in 539a10).

operation of the earlier part, but rather because the generative, innate principle of motion continues its sequence of actualizations, by comparing the generation of parts to the operation of automatic puppets.[14] Aristotle points out though (in 734b27–28) that the generation of the homogeneous parts such as blood and flesh and of the instrumental ones such as hands and feet come into being *at the same time* (ἅμα δὲ τὰ ὁμοιομερῆ γίγνεται καὶ τὰ ὀργανικά), since the latter cannot exist without the former. He also makes clear that there is one part that comes into being first and "not at all at the same time" (735a15: οὐχ ἅμα πάντα) and that it has within it a principle of growth, and this is the heart.

The question of order is of course one that is of intrinsic interest to Aristotle's discussion of embryogenesis, but it also reveals concerns that are congenial to Aristotle's philosophy of science as recorded in his *Posterior Analytics* and to his attempts to track to the order of nature in the order of exposition. Granted, this connection with the theory of demonstration outlined in the *APo* is not explicit here, but the terms that Aristotle uses in this section to discuss the coming to be of parts – does it happen "at the same time" (ἅμα) or consecutively (ἐφεξῆς) – are exactly the same ones he uses in *APo* 2.12, where he describes in a more technical context three different models of demonstrations of natural processes, each of which tracks the chronological order of causation in the world. The third of these models involves circular coming-into-beings, which are not relevant here, but the first and the second model differentiate between cases where (1) an attribute comes to hold of a certain subject at the same time as the event picked out by the middle term occurs, such that cause and effect occur simultaneously (ἅμα),[15] and cases where (2) the items picked out by the three terms do not occur at the same time but consecutively, such that first this happens, then that.[16] It is thus important to identify the exact

[14] *GA* 2.1, 734b9–16: "It is possible that the one moves the other, and the other another one, and that is the same as with automatic puppets. For their parts while at rest have some kind of potentiality: when something external moves the first of them, immediately the next one turns into actuality." Cf. *GA* 2.5, 741b7–9: "Since the parts already exist in potentiality in the matter, as soon as the principle of motion comes to be they get strung together *in order* (συνείρεται τὸ ἐφεξῆς) just as in the case of automatic puppets."

[15] *APo* 2.12, 95a10–24 (a22: "τὸ αἴτιον καὶ οὗ αἴτιον ἅμα γίνεται").
Example: A = eclipse ["eclips*ing*"] A = solidified
B = earth *coming* in the middle B = complete *cessation* of heat
C = moon C = water
The coming to be of A(aC) $\approx_{sim}$ coming to be of B(aC).

[16] *APo* 2.12, 95a24–95b1 and *APo* 2.12, 95b13–b37 (95b13: ἐφεξῆς γινομένης); cf. *GC* 2.11, 337b14–25.
Example: A (stones) at t_1 → C (foundation) at t_2 → D (house) at t_3

sequence between events – whether they happen simultaneously or consecutively – because different types of events require different models of demonstration and they each have different requirements for how the premises get tensed, etc.

In this way, the concerns with providing the materials for demonstration and the attempt to track whatever the relevant natural order is in exposition, go hand in hand: since the account in the *GA* is pre-demonstrative (by which I mean that Aristotle aims to provide causal explanations of the natural phenomena[17] that could perhaps be fleshed out in scientific demonstrations in the strict sense, but that are certainly not presented as such), and since demonstrations are supposed to track the order of causality in the world, Aristotle's exposition will have the most explanatory force if it reflects the order of coming to be of parts in his writing. And it is in fact the very precise order in which parts come to be during the development of blooded, live-bearing animals such as in particular humans that guides Aristotle's exposition in *GA* 2.4–6, as we will see in the next section of this chapter.

The third puzzle that Aristotle addresses in the opening chapters of book 2 and that has relevance for his concern with order pertains to the question of what happens to the material part of the semen through which the generative principle is imparted into the female (*GA* 2.3, 736a24–737b6). As it turns out, answering this question involves not just determining whether the female receives anything material from the male semen (here Aristotle's answer is clearly no), but also whether the sensitive soul is already in the male semen and in the embryo, and if so, where this sensitive soul comes from. The full answer ends up requiring a discussion of when and from where exactly the embryo receives the different kinds of soul. The exposition here is difficult and the text is in part corrupt, but for our purposes it suffices to see that Aristotle conceives of the reception of soul as a three-stage process, with the embryo reaching a more perfect level of being with each acquisition:

(1) The nutritive soul the embryo possesses from the beginning in potentiality (736a35–b1; b8–13); it is present in potentiality in the female menses. Having this soul turns the embryo into something plant-like.
(2) The sensitive soul the embryo acquires later in potentiality through the male semen that possesses vital heat. Having this soul turns the plant-like developing being into an animal embryo.

[17] See, e.g., *HA* 1.6, 491a7–13 for Aristotle's own statement of purpose.

(3) The rational soul poses the biggest difficulty of all: it is divine and enters the embryo "from outside." Having this soul turns the animal embryo into a human one.

Again, the discussion here is interlaced with references to what happens "at the same time" versus what happens earlier or later "consecutively"[18] and the order of exposition traces the order of the acquisition of the different souls, which slowly "transforms" the embryo and the developing offspring into a human being.[19]

Having settled these puzzles, Aristotle can now turn to his discussion of the process of embryogenesis properly speaking.

Discovering the Order of Embryogenesis of Live-Bearing Animals in *GA* 2.4–6

In this section, I turn to a third sublevel of organization that pertains to chapters 2.4–6,[20] which covers the process of embryogenesis in the most perfect animals, that is, in humans and other live-bearing animals. Note that what we get in these chapters is indeed a self-contained "subargument" is clear from both the opening of *GA* 2.4 and the opening of *GA* 3.1. For after Aristotle's brief summary at the opening of 2.4 of the three kinds of processes involved in reproduction, i.e., live-bearing, egg-laying, and grub-producing, he announces that it is time to make a *first beginning* with a discussion of the first beings (*GA* 2.4, 737b25: νῦν δ' ἀπὸ τῶν πρώτων ἀρκτέον πρῶτον), which are explicated to be human beings and other live-bearing animals. The opening of *GA* 3.1 claims that the live-bearers have now been discussed, and turns to embryogenesis in egg-layers as its next topic. Within these intervening chapters, Aristotle reverts back to using the chronological order of the process of embryogenesis as the principle of organization, but faces some methodological challenges in trying to discover what exactly this order is.

[18] See e.g. *GA* 2.3, 736b1–4: "For one does not become animal and human at the same time (οὐ γὰρ ἅμα γίγνεται), nor animal and horse, and similarly with regard to the other animals. For the end is the last to come to be (ὕστατον γὰρ γίγνεται τὸ τέλος), and what is specific is the end of each generation."

[19] Note that Aristotle uses the same sequence of souls in his function-argument (*NE* 1.7, 1097b33–1098a4): after the nutritive capacity has been rejected as being constitutive of the human function, the following (1098a1: ἑπομένη) capacity that needs to be considered is the perceptive one, and once that one is rejected, the only capacity that is "obviously left" (1098a3: λείπεται δή – I am overtranslating δή here) is the rational one.

[20] Note that *GA* 2.5 and *GA* 2.7–8 deal with additional puzzles rather than with the process of embryogenesis strictly speaking and so I will have very little to say about these chapters in my account below.

Aristotle divides his discussion of the formation of embryos in live-bearing animals into two parts: the development up until the formation of the heart, and the development after the heart has taken shape. The description of each of these phases comes with its own methodological challenges. The first stage consists of the formation of the homogeneous reproductive parts: Aristotle first discusses the secretion of the (more perfect, male) semen (in *GA* 2.4, 737b27–38a9), then the production of female menses (in 738a9–39a6). The second stage consists in the formation of the embryo: as soon as the male semen enters the uterus, it organizes or "fixes" (συνίστησι) the purest part of menses (739a6–39b20). Once the menses have been "fixed" by the semen (this entails bringing together solid parts and separating of liquid parts), membranes form through the solidification of earthy parts (739b20–b33). And then, once the embryo (τὸ κύημα) is "fixed," it acts almost like the seed of plants, and sends off "the shoot and the root" (i.e., two blood-vessels from the heart, from which smaller vessels branch off to the uterus and which form the umbilical cord: Aristotle explains this more clearly in 740a24–35), through which it receives nourishment for growth (739b33–40a1). The third stage involves the formation of the parts of the animal, beginning with the formation of the heart (740a1–24). As Aristotle explains, the embryo possesses all parts in potentiality, but the "first principle is the furthest on the road [toward realization]" and so it comes into actuality first.

In other words, according to Aristotle the heart is visually the first part that can be distinguished in the developing embryo (740a17–18), and the other parts take its principle of growth and organization from it, while getting their material or nutrition via the umbilical cord. At this stage, the embryo is basically just the heart encapsulated in a membrane with the purest part of the female menses fixed inside of it and with vessels branching off of it; Aristotle characterizes it as "potentially an animal, but an imperfect one." He concludes the chapter by identifying the agent of the growth of the embryo as the nutritive soul, which is identical to the generative soul, or the nature that is present in each of the different kinds of living beings (*GA* 2.4, 740b29–741a2).[21]

Two methodological issues stand out in Aristotle's discussion of these early stages of embryogenesis. The first is raised explicitly by Aristotle and

[21] *GA* 2.4, 740b36–741a2: "So if this is the nutritive soul, it is also the generative soul; and this is the nature that is present in each thing (καὶ τοῦτ' ἔστιν ἡ φύσις ἡ ἑκάστου ἐνυπάρχουσα), both in plants and in animals; both the other parts of the soul belong to some, but not to others of the living beings."

pertains to how exactly one establishes which part comes to be first. For that it is the heart that is formed first in actuality during embryology is apparently a controversial point, and Aristotle argues for it by pointing both to perceptual evidence (740a4: ἐπὶ τῆς αἰσθήσεως δῆλον) and to theoretical considerations (740a5: ἐπὶ τοῦ λόγου). The argument goes something like this: (premise 1) the organizing principle must be present in what is being organized from the start; (premise 2) blood or its analogue is the nutriment of the body; and (premise 3) since the heart is the receptacle of blood or its analogue; therefore (conclusion) the heart is the principle of growth for all the homogeneous and heterogeneous parts and must be present first. Aristotle criticizes people like Democritus[22] who claim that "the outer parts are differentiated first, and later the internal ones," for they "do not speak correctly, as if they were speaking about wooden or stone animals. For such [animals] do not have a principle at all, but animals all have it and have it internally." Note that in the next chapter (*GA* 2.5, 741b15–24), Aristotle appears to be adding another reason for why it is the heart that must come to be first. He reiterates that all the parts are present in potentiality in the embryo, that they develop in a chain, one after the other, and that it is the heart that comes to be first as is clear to perception, but then adds a kind of teleological consideration: as a general, natural rule, something that comes to be last, fails first, and what comes to be first, fails last, and since it is the heart that fails last when someone dies, it must have come to be first. This seems to be connected to Aristotle's idea – about which I will say more below – that the more essential or vital a part is, the earlier it must come to be, whereas parts that come to be last must be the least essential or vital.

The second issue that is of methodological interest, and which is more salient to us as a contemporary audience, is that Aristotle relies heavily on analogies in order to provide a detailed description of the early development of the embryo. For until the heart is present and can be marked out by observation (during dissections of animals), no other part of the process is readily available to observation. Because of this, Aristotle draws from well-known[23] organic processes to visualize – and causally explain – what must go on in the early formative stages of the embryo. For instance, Aristotle claims that the male semen – or, more precisely, the active principle in it – "fixes" the female menses *in almost the same way* as rennet

[22] And potentially also some Hippocratics. Cf. Hippocrates, *On the Nature of the Child* 17.2–3.

[23] The cheese making and botanical analogies Aristotle uses are both very common in the Hippocratic treatises on embryogenesis as well.

acts upon milk (739b21–22: παραπλήσιον ποιούσης), as both are said to contain vital heat, and menses and milk are said to share the same kind of fluid nature with earthier and purer parts included. And in order to explain the growth of the two blood-vessels from which also the umbilical cord is separated off, Aristotle compares the growing embryo to the seeds of plants: "it acts almost (παραπλήσιον ποιεῖ) like seeds. For seeds also contain the first principle of growth in themselves, and when this has been differentiated [. . .], the shoot and the root are sent off from it. And it is by the latter that it gets nourishment; for the plant needs growth" (*GA* 2.4, 739b33–740a1; the analogy extends all the way through 740b2; cf. 740a34: οἷον ῥίζαι). In these cases, Aristotle relies on more familiar and visually accessible phenomena in order to provide a chronological account of the development of the embryo.

After a brief detour offering a solution for the puzzle of why females cannot generate on their own, Aristotle returns to his discussion of the order in which the parts of live-bearing animals come to be: the internal parts come to be before the external, and the parts above the diaphragm come to be before the ones below. This latter claim is as much based on observation for Aristotle as it is on a teleological "ordering" principle that he had mentioned earlier in the context of his discussion of why all females have menses but not all males have semen (*GA* 2.4, 738b15–18): "This is the region about the diaphragm in all those animals which have one, for the heart or its analogue is a principle of nature, and the lower part is added to it and is for it" (τὸ δὲ κάτω προσθήκη καὶ τούτου χάριν). The idea that the above is better or more honorable, as Aristotle likes to put it, than below, right better than left, and front better than back is common in Aristotle's biology.[24]

However, as Aristotle makes clear from the outset, determining exactly in which order which part comes to be after which is complicated, because observation is not necessarily reliable in all these cases: bigger parts become visible before smaller parts even if they did not come to be before them (741b26–27). And so Aristotle criticizes his predecessors for lacking the appropriate experience and sets out to develop his own method for determining which part comes to be after which.

Schematically, developing this method involves a conceptual analysis of what it means for something to be "prior in nature" and subsequently of

[24] See, e.g., *PA* 3.4, 665b18–21; *PA* 2.14, 658a18–24; and *IA* 2, 704b18–22. Cf. also *GA* 1.8, 718b25–29 and 1.11, 719a13–15.

correctly identifying the causal role that each part plays in the animal's body (*GA* 2.6, 742a16–742b17):

(1) One needs to distinguish between "that for the sake of which," which is prior in substantial being, and "that which is for the sake of it," which is prior in generation;[25]
(2) One needs to distinguish in the category of "that which is for the sake of it" between that which has the source of motion (or what has the generative capacity or is called the producer) and that which is used by that for the sake of which, and which is thus instrumental;
(3) From these two distinctions, one learns that there are thus three kinds of things, each characterized by its own causal role and with its own spot in the temporal order of coming to be:
    (a) the end or that for the sake of which (such as for instance the learner), which in the temporal order of coming to be is second;
    (b) the source of motion or generative capacity (such as for instance the teacher), which in the temporal order of coming to be is first; and
    (c) what is useful and what the end makes use of (such as for instance the flute), and which in the temporal order of coming to be is third.
(4) Applying this to biological parts, Aristotle concludes that there must be:
    (i) One part that is the source of motion;
    (ii) "The whole and the end," which is presumably the complete body;
    (iii) The parts that are instrumental to the whole and the end;
        (iii.a) parts that are productive (i.e. involved with the generative process) must be present not first, but at a very early stage;
        (iii.b) parts that are not productive but for the sake of something else must come to be later.

Aristotle admits that it is not easy to tell which parts are prior and which are for the sake of which, but states that nevertheless "it is necessary to search for what comes to be after what in accordance with this method" (742b10–11: καίτοι κατὰ ταύτην τὴν μέθοδον δεῖ ζητεῖν τί γίγνεται μετὰ τί). With this, Aristotle briefly criticizes those like Democritus who deny that it is possible

[25] On this distinction between what is "prior in nature"/"prior in substantial being" and what is "prior in generation," see *PA* 2.1, 646a24–b9 (building on *PA* 2.1, 640a10–b4): "Since in generation things are opposed to the way they are in being – for things posterior in generation are prior in nature, and what is first in nature is last in generation. For a house is not for the sake of bricks and stones, but those are for the sake of a house. [. . .] It is thus necessary that the matter and generation are prior in time, but in definition the being and the shape of each thing."

to find the first principle responsible for the growth of animals, and reiterates that in those that come to be, one of the first principles is the source of motion, and therefore all animals develop first the heart or its analogue (*GA* 2.6, 742b18–37).

With this method in place, Aristotle continues to describe the process of embryogenesis in chronological order by pointing first to anatomical diagrams on the wall (743a2: γράφοντες ἐν τοῖς τοίχοις), which represent the blood-vessels stretching out from the heart, with the parts around them growing out of these vessels. He also explains the mechanisms of heat and cold that nature (or the nutritive, generative soul) uses to produce the homogeneous parts such as flesh and bones that fill up the body so to speak (743a1–743b18). Next, using a painting analogy, Aristotle states that the upper half of the body "is marked out (ἀφορίζεται) first in the order of development," the lower half reaches its full size later, and that all the parts are "first drawn in their outlines (ταῖς περιγραφαῖς διορίζεται)" and only later take their permanent color and softness or hardness, as if nature is first "sketching in (ὑπογράψαντες)" the animal (743b18–24).[26]

Now, with more precision as to what part comes to be after what, Aristotle returns to the heart, which is also the seat of sensation. Because of the heat produced by this part, the cold produces the brain next, as the part that corresponds to the heart. The other parts on the head start to grow, and not just the brain, and soon they surpass the heart in size (743a25–32). Then a difficulty follows (743a33–744b11): one of the sets of parts on the head, the eyes, comes to be early, but they are the last to complete, and this requires an explanation (which Aristotle provides in terms of both necessity and teleology: it takes time for eyes to receive their constitutive matter and since nature does nothing in vain the eyelids should not be separated until the heart is able to move them). In order to give the chronology of the coming to be of the *remaining* parts, Aristotle finally appeals to his own method of determining the causal roles of parts first to determine their order in coming to be and he does so while using an elaborate analogy between nature and a good housekeeper. Let me quote the text in full:

[26] Sometimes Aristotle depicts the actions of nature itself as those of a sketcher or painter: nature is said to have "sketched in," i.e., ὑπέγραψε, skeletal analogues in sepia and squids in *PA* 2.8, 654a24–6 and hair in the fronts of humans in *PA* 2.14, 658a21–3; in *PA* 2.14, 658a31–5 nature is said to have "decorated," i.e., ἐπικεκόσμηκεν, long tails with hair; and, earlier in *GA* 2.4, 740a28–9, we saw that Aristotle says nature "sketched in" (ὑπέγραψεν) two blood-vessels running from the heart. Cf. the comparison of flesh to a frame-work (ὡς περὶ ὑπογραφήν) in *GA* 4.1, 764b30.

> Each of the other parts is formed out of the nutriment, (A) the parts that are the noblest and that partake in the most important principle are formed from the nutriment which is concocted first and is purest; (B) the parts that are necessary, that is to say that are for the sake of the former parts, are formed from the inferior nutriment and the residues and leftovers. For just like a good housekeeper, so also nature is not in the habit of throwing away anything from which it is possible to make anything useful. Now in a household the best part of the food that comes in is set apart for (A) the free people, the inferior and the residue [of the best food] for (B1) the slaves, and the worst is given to (B2) the animals that live with them. Just as the intellect from the outside[27] does those things with a view to growth, so nature in the things coming to be forms from the purest material (A) the flesh and the body of the other sense-organs, and from the residues thereof (B1) bones and sinews and (B2) hair, and also nails and hoofs and all similar parts; for this reason these are the last to assume their formation, for they have to wait till the time when nature has some residue to spare. (*GA* 2.6, 744b11–745b15)

Note that Aristotle divides the remaining parts into two groups: group A, which presumably includes the parts that are "that for the sake of which" and are of vital and essential importance to the animal, and group B, which presumably includes the parts that are for the sake of something else. Now in a household, the hierarchy is not two- but tri-fold, and those at the top get fed the best food, and get fed first, while those below get the leftovers and get fed later. In the same way, or so Aristotle argues, does nature produce the live-bearing animal: first, the best parts – such as flesh and the sense-organs – are made from the best matter available, and from what is left from this, the lesser parts – such as bones and hair – are formed.

Aristotle then continues to specify this picture as follows: while nature is producing the noblest parts in group A (flesh and sense-organs) during the "first formation," it also produces bones, but the bones come to be from the spermatic residues that are left after the parts in group A are fully formed. And just as the coming to be of bones is secondary upon that of flesh and of the sense-organs, so is their growth: bones take their growth from the – lesser – residues of the natural, nutritious food that is not used up by the flesh and the sense-organs (744b28–36). The coming to be of sinews parallels that of bones – they also come to be second during the first formation (744b36–38). Together these parts appear to make up what I have labeled as group B1 in the text. The other parts, labeled as group B2 above, come to be later (745a1–18): as Aristotle explains, parts like nails, hair, hoofs, horns, beaks, spurs of cocks, and other similar parts are made

[27] For the *crux* in the text, I here follow the conjecture by Moraux (1955: 255–95).

nutriment that is taken in *later* and that is only concerned with growth: the materials constitutive of these parts are said to be "derived from the mother or from the outer world after birth." Aristotle uses this distinction in the constitutive matter to explain why bones cannot keep on growing (there is a limit to the amount of residues of the natural, nutritious food, and also a formal limit to their size), whereas hair can – as long as there is residual matter from food taken in later in life, hair can keep growing.

Aristotle concludes with a puzzle about teeth, pertaining to their categorization (745a19; 745b2–4): are teeth like bones (and sinews) or like hair and nails? Teeth fall in the class of parts that are instrumental for Aristotle,[28] so clearly in group B, but do they belong to B1 or to B2? As Aristotle points out in the next lines, teeth are of the same nature as bones (cf. *HA* 1.11, 493a2), and thus belong to B1, but not without qualification: the coming to be of teeth is intrinsically connected to the growth of bones, but in a way that makes their coming to be dependent on and secondary in time to the growth of bones (*GA* 2.6, 745b2–5): "And it turns out that teeth do not have the same nature as the other bones: for they all come to be in the first formation and none of them later (τὰ μὲν γὰρ ἐν τῇ πρώτῃ συστάσει γίγνεται πάντα καὶ οὐθὲν ὕστερον), whereas teeth come to be later." The explanation lies in the material from which teeth come to be: teeth come to be from the food distributed to the bones but that is not used up by the bones for their growth. Once the bones "have reached their number," teeth come to be (*GA* 2.6, 745b7–9). This particular connection of teeth to the growth of bones also accounts for two of their distinctive features. First, unlike the other bones, teeth can keep growing throughout the animal's lifetime (*GA* 2.6, 745a25), as long as there are residues available from the food distributed to the bones. (As Aristotle is quick to point out in 745a27–b2, the continuous growth of teeth also serves a final cause: they would blunt fast, if there were not some flow of materials toward them.) Second, because teeth are not grown together with bones and come to be later, they can also grow back again (only once in human beings, but repeatedly over their lifetime in some of the other animals with teeth, such as in sharks and reptiles) after they have fallen out.

Aristotle ends by stating that because parts like hair and nails are formed from residues, humankind is most naked of all animals and has the smallest nails in proportion to his size, for he has the smallest amount of earthy

[28] Teeth are present for the sake of nutrition, and sometimes also for defense and speech; see *PA* 2.9, 655b2–15; *PA* 3.1, 661a34–662a15; *PA* 4.5, 678b6–26; cf. also *PA* 3.2, 663b22–35, and *PA* 3.14, 674a12–22.

residue (*GA* 2.6, 745b16–20). And with this remark Aristotle claims that he has now stated how *each* of the parts is constituted, and what the cause of their generation is. Now obviously, Aristotle did not mention every single part in the human body individually, but I take it that since he covered each of the three "groups" of parts (the source, the parts that are that for the sake of which, and the parts that are for the sake of the latter, of which some come to be during the "first formation," others later and even after birth) and explained each part about which there exists a difficulty (namely eyes and teeth), his account is now considered complete: for each individual part not named, figure out what causal role it plays in the body, and "searching in accordance with this method," you will find out at what point during embryogenesis it comes to be.[29]

Concluding Remarks

Aristotle's *GA* – including its second book – is a highly organized and coherent text: at the most general organizational level, its structure tracks the chronological order of the process in which animals come to be. Aristotle thus starts with the identification and definition of the first principles of reproduction and ends with an account of the differences of attributes that come to be post birth and sometimes even later in life. The secondary level of organization – or the second argument, so to speak, that is embedded in the "one long" argument that incorporates the whole of *GA*, and which pertains to chapters 2.4–3.11, tracks the natural order of perfection among different animal kinds. In these chapters, Aristotle discusses the distinctive process of embryogenesis in each particular animal kind, starting with an account of the process of embryogenesis that leads to the most perfect offspring in the most perfect kind of animal, and ending with an account of spontaneous generation in the least perfect kind of animal. And finally, within chapters 2.4–2.6, Aristotle reverts to following the chronological order of embryogenesis again, but this time specifically in the most perfect of animals and at the most detailed level possible, starting with the coming to be of the embryo and the heart itself, and ending with the parts that come to be later and sometimes after birth, such as teeth and hair. Tracking the various orders of nature takes effort, and Aristotle takes great pains to develop the appropriate methodological tools to achieve this.

[29] For instance, kidneys presumably come to be during the first formation but second in order of coming to be after the bladder has come to be: kidneys are instrumental but not generative (see *PA* 3.7, 670b23–27) and therefore come to be later.

PART II

The Principles of Animal Generation Reconsidered

CHAPTER 4

What Is Aristotle's Generation of Animals *About?*

Pierre Pellegrin

Introduction

Reproduction is, according to Aristotle, a function proper to living beings, at least to those located in the sublunary region (the celestial bodies and the prime mover are described as living beings by Aristotle, but they do not reproduce). In fact, reproduction in its proper sense applies only to animals.[1] Aristotle does, as we do, consider an animal as an integrated system of great functions with certain important differences. Even if they use similar or approaching terms, Aristotelian and modern descriptions do not refer to the same things. For instance, in Aristotle's view, respiration is a way of cooling the blood, as he does not know the distinction between arterial and venal blood and, moreover, he does not think of blood as *circulating* in the sense of the same blood coming back in the vessels after having been cooled in the lungs.

We clearly see that not only the functions of the same organs are not identical in Aristotle and in modern physiology, but also that the relations between the different functions are not the same. According to the modern view, a mammal such as a human being, or close enough to a human being, has at least the following functions: a skeleton made of bones and cartilage (which, strictly speaking, is not a function), a muscular and motive system, a respiratory system, a circulatory system, a digestive system, a reproductive system, and a nervous system. To describe an animal as an integrated system of functions means that these functions have some interaction. Respiration, for example, is closely linked to circulation, since blood gets some oxygen in the lungs. These functions are also closely related in Aristotle's physiology, since the latter posits that blood is cooled in the lungs. This is precisely what Aristotle means when he says that the entire body has "total action" that is the sum of the particular actions of the

[1] The reproduction of plants is a separate problem I want to leave aside here.

particular parts (*PA* 1.5, 645b14–20). Yet, this doctrine is completely different in its content in modern and Aristotle's cases. There is nothing in Aristotle's physiology that may correspond to the nervous system: the brain has no psychic function; rather, it has a cooling function, counterbalancing the heating effect of the heart. Moreover, the "nerves" are nothing but sinews moving limbs as ropes would do. In his seminal article *Greek Philosophy and the Discovery of the Nerves*, Friedrich Solmsen shows that Aristotle is in an intermediate position: he has abandoned the doctrine of blood as being a medium for sense-perception and motion (a doctrine adopted by the Sicilian medical and philosophical school originating with Empedocles and more or less adopted also by Plato in the *Timaeus*) but does not yet consider nerves as transmitters. His doctrine of the connate *pneuma* would be an evidence of such a middle position.[2] Nor does Aristotle have the notion of muscles. He just speaks of *flesh*, the motions of the limbs being activated by sinews (νεῦρα).

I would like to make an additional point, one which may seem external to the subject I am dealing with but which could help us grasp more precisely Aristotle's position vis-à-vis what we today call biological and medical sciences. When writing physiology, in his biological corpus and elsewhere, Aristotle obviously borrows some notions from the medical tradition. Marwan Rashed has made this quite clear in the case of the treatise *On Generation and Corruption*.[3] In fact, Aristotle is one of the Greek philosophers who also belongs to the history of medicine. The close connection between philosophy and medicine was very common among the Presocratic philosophers (Empedocles, Parmenides, Heraclitus) but came to an end with Socrates. It was revived for the last time by Aristotle (so long as we set aside the obscure and, so to speak, "negative" relationship of empiricist and methodist medicine to skeptic philosophy). This was recognized by medical writers already in antiquity – for instance, by the leading figures of Alexandrian medicine like Herophilus and Erasistratus, and by Galen five centuries

[2] Solmsen 1961. It would be quite interesting to show in more details how the *pneuma*-bearing vessels have, so to speak, transformed themselves into nerves. Here I am content to say that Praxagoras of Cos played a crucial role in this transformation. Praxagoras adopted the doctrine of the *pneuma*, which was carried by arteries (he is often thought to be the first to distinguish between veins and arteries, but this cannot be entirely true, since such a distinction can be found already in Aristotle). These arteries become smaller and smaller, to the point that they end up transmitting something that is almost immaterial. At this point, Praxagoras calls them *neura*. It was Praxagoras' pupil, Herophilus of Chalcedon, who first distinguished sensitive and motive nerves, and it was his junior colleague Erasistratus of Ceos who developed this doctrine further. This takes us well into the third century BC. Cf. Steckerl 1958 and von Staden 1989.

[3] In the introduction to his edition of the *GC*. See Rashed 2005: xxiv–xxx.

later, who adopted explanatory systems which were basically Aristotelian. The Stoics, and even more so the Epicureans, on the other hand, *alongside their philosophy*, have recourse to some medical doctrines, particularly when speculating on the topic of the soul.[4] We should, of course, recognize some influence of Stoicism on some physicians of the so-called "pneumatic" school of medicine (most notably Athenaeus of Attalia, who seems to have inspired rather than founded the school, and who probably was a Stoic), but this Stoic influence is in no way comparable to that of Aristotle on ancient medicine.[5]

So, after the biological turn, which has been so influential among the Aristotelian scholars, should we be prepared to face a "medical turn," medicine being a key to the understanding of Aristotelian *philosophy*? Perhaps so. In any case, Aristotle is the last ancient philosopher who made what we call physiology a chapter of his philosophy.

The *GA* as a Treatise on Generation and Nutrition

Several passages in the Aristotelian corpus refer to a treatise which is, beyond any doubt, Aristotle's *GA*. These references are sometimes to an existing work, sometimes to a work to be written. In all these cases, Aristotle refers to a treatise in which the different modes of animal reproduction are or will be considered, including an examination of the reproductive organs, the nature of semen (σπέρμα), and various questions concerning the offspring. The *GA* exemplifies, more than any other Aristotelian work, a very typical kind of speculation: a *philosophical* investigation (and as such based on principles of Aristotelian philosophy, like matter and form, potentiality and actuality) dealing with questions that belong to physiology and as such will be considered by later physicians.[6] But, at this point, we should notice that when referring to a treatise on the generation of the animals – whether these references can be clearly identified in the work on the generation of animals we actually find in the Aristotelian corpus or not – Aristotle does not only assign to this treatise the study of the function of reproduction but also that of nutrition.[7]

[4] Cf. Pellegrin 2006.

[5] The pneumatic school eventually adopted an Aristotelian approach. But this is another story.

[6] See, for instance, Galen's *On Semen* [*De semine*]. In this work, Galen deals with the development of the fetus and proves, against Aristotle, that the substance of the semen provides some matter for the fetus.

[7] Here are some cross-references. *PA* 2.3, 650b9–10: "The way in which the parts grow from blood, and also the subject of nourishment generally speaking, is more appropriately considered in the works on generation as well as in other works"; *PA* 3.5, 668a6: "As to how and from what animals are nourished, and in what way they take in nourishment from the gut, it is more appropriate to

This is perhaps the most striking example of the already mentioned difference between Aristotle and modern biology on the demarcation of the different physiological functions.

Let us begin by quoting two passages from the *PA*:

> It is necessary that the incoming nourishment be received and that the dehydrated nourishment be expelled; and the unconcocted nourishment and the residue must not be in the same place, and there must be a certain place in which the nourishment is transformed. For the one part will hold the incoming nourishment, the other the useless residue; and as it is necessary for each of these operations to take place in a distinct time, it is also necessary that they be divided in their place as well. *But since the definition of these parts is more appropriate to the works on generation and on nutrition* [. . .]. (*PA* 3.14, 674a13–21)[8]

> What the mesentery is for, then, has been said. How it takes up nutriment and how what is distributed to the blood vessels comes into these parts through the blood vessels from in the incoming food, *this will be said in our works on generation and on nutrition.* How the blooded animals are designed as far as the parts already defined are concerned, and according to what causes has been said. Next, among things to be spoken of are the parts that contribute to generation by which we believe that female differs from the male. But since we need to speak about generation, it is appropriate to consider the generative parts as well in the study of those things. (*PA* 4.4, 678a15–25)

Some years ago, I was tempted to believe that the references to the lost Aristotelian work *On Nutrition* were in fact references to the *GA*, or to some part of the *GA*. But this is not likely to be true. For when in *GA* 5.4, 784b2–3, Aristotle says "we shall speak more in detail about this cause later in the work on growth and nutrition," this can hardly be an internal reference.

This last passage shows that the difference between Aristotelian and modern points of view is even greater than we may initially think. For, according to Aristotle, the preservation of the body by nutrition, the growth of animals, and their reproduction belong to one and the same study. Of course, nutrition and reproduction are two different activities,[9] but they both depend on the same power (δύναμις): the nutritive soul

investigate and to speak about such things in the works on generation"; *GA* 2.1, 735a17: "The nutritive power exists in all alike, in animals as well as in plants, and this is the same power which enables an animal or a plant to generate another being like itself."

8 All translations are my own unless otherwise indicated.

9 Cf. *PA* 2.7, 653b13–18: "The residues of nutriment should be studied in the examination and study of nourishment, stating in which animals they are present, and for what causes, whereas the residues of semen and milk should be studied in the works on generation; for the first of these residues is the principle of generation, and the second is for the sake of generation."

(θρεπτική ψυχή). The nutritive soul is the power that makes an animal able to assimilate food for the sake of the accomplishment of the three aforementioned functions, namely preservation, growth, and reproduction.

That these functions are different, even if they depend on the same power, is made clear by a question Aristotle considers, namely that of the order to be found among these powers. For the moment, let us consider under the denomination "nutrition" both preservation and growth, which will be distinguished later. We then have Aristotle's position clearly expressed in the treatise *On the Soul*:

> The nutritive soul belongs to all living beings, and it is the most primitive and the most common function of the soul, being that in virtue of which all are said to possess life. The actions in which it manifests itself are reproduction and the use of nutriment. (*DA* 2.4, 415a23–26)

But concerning the priority of one these two functions Aristotle adopts two positions which seem to be inconsistent to each other. This may be seen in two passages from the same chapter from the work *On the Soul*. The first is this:

> Since it is right to call things after their goal, and since the goal is to generate a being similar to itself, the first soul will be the generative principle of such a being. (*DA* 2.4, 416b23–25)

This passage seems to make generation prior to nutrition. However, a few lines earlier we are told this:

> Since nutrition and reproduction are due to the same power of the soul, it is necessary to deal with nutrition first, for it is by this function that this power of the soul is distinguished from all the others. (*DA* 2.4, 416a18–21)

We do not have here the well-known Aristotelian distinction between what is prior by nature and what is prior to us, but rather a distinction between what is prior because it is a goal and what is prior because it is more basic. But why are the nutritive soul and the generative soul one and the same thing, at least to some extent? It is hard to resist the temptation to read this doctrine through a Freudian lens. For it is a characteristic feature of the theories children elaborate to solve the riddle of their birth, relate pregnancy to the ingestion of food, and make delivery happen through the anus. If they were able to write, children would also write treatises on generation and nutrition. I won't adopt such an explanatory posture, restricting myself to a comprehensive one. That Aristotle ascribes the nutritive and reproductive function to the same soul is not, of course, a discovery of my own. But my impression is that until now no one has explained why it is so, or at least has explained it completely.

There is a first, weaker reason why the *GA* is a treatise on nutrition and generation, which is that the elements of generation (i.e., semen and menstrual matter) are simply elaborated food, with blood as an intermediate state. So among other things, when nutriment is entirely used for growth in young animals, they do not produce any semen because there is no extra nutriment left to do so. From the point of view of this weaker reason, which is a *valid* one, the doctrine of the elaboration of nutriment into blood and subsequently into semen belongs to the study of animal generation. It should be noted that this weaker reason is the only one interpreters offer to explain the fact that Aristotle considers the nutritive soul to be also generative. Bernard Besnier, for instance, does so in one of the very few articles entirely devoted to the topic of the nutritive soul in Aristotle.[10] In fact, this article is mainly concerned with the fact that the nutritive soul has to do with preservation and growth, which is not a small step in the right direction.

But there is another, stronger reason why the three aforementioned functions (preservation, growth, and reproduction) depend on the same power of the soul, and more precisely on the nutritive soul. To see this, we should consider each of these functions in more detail. We can start by noting that, according to Aristotle, there are three motions within an animal that are natural, namely that are not mainly produced from outside. Since they are the same from the point of view of *matter*, they can be grasped according to the same schema but are nevertheless different in *being*. The first is the preservation of the organism by nutriment. This motion is mainly described as the transformation of nutriment into blood and of blood into flesh. Aristotle does not say much on the formation of the earthy parts like bones, cartilage, nails, and horns. He just says that one kind of earthy matter goes upward, while the other goes downward. This first motion is a case of alteration (ἀλλοίωσις). The second motion is the growth of the animal (αὔξησις). This motion is described as an assimilation of nutriment by the organism and not as a mere addition to it. The generation (γένεσις) of the animal is the third and final motion. It is clear that generation and growth overlap, since growth begins *in utero*, a process that the *GA* describes in some details in *GA* 2.6. The description of these three processes is to be found in the *PA*, in the *GA*, and more completely in the *GC*.

[10] Besnier 1997.

Let us recall a couple of passages about this triptych. That growth and generation are the same thing is explained, among other passages, here:

> The female provides matter, the male the principle of movement. And as the products of art are made by means of the tools of the artist, or more precisely by means of their movement, and this movement is the activity of the art, and the art is the form of what is made in something else, so is it for the power of the nutritive soul. As this soul produces later growth from nutriment for animals and plants, using heat and cold as its tools (for in these is its movement [of the power of the nutritive soul], and each living being is produced according to a definite ratio), in the same way from the beginning does it form what is generated by nature [sc. the embryo]. For the matter by which the latter grows is the same as the matter from which it is constituted at first; consequently (ὥστε) the productive power is also identical from the beginning. But the latter [sc. the power that produces the embryo] is greater. Thus if it is the nutritive soul, it is also the generative soul. This [sc. the nutritive soul] is the nature of every organism, existing in all animals and plants. (*GA* 2.4, 740b24–741a3)[11]

This text could be endlessly discussed. What is here called the generative power of the nutritive soul[12] is the organogenetic faculty that the living being possesses in itself once the principle of the embryo has been formed. This principle, viz. the heart, is the result of the action of the motions conveyed by the semen on the female matter, which Aristotle compares to the action of rennet upon milk (739b21). In order to, so to speak, underline the threptic character of this generative function, Aristotle compares this process to the growth of a plant through its roots: just as the heart produces blood-vessels which are like roots (740a33), the growth of the embryo is something like the growth of plants through their roots (740b10). This last passage is crucial for my interpretation, since Aristotle says that the formation of the embryo, the growth of plants, and the growth of animals already born happen "in the same way" (τὸν αὐτὸν τρόπον, 740b9).

It is difficult to decide what force is to be given to the inference beginning with ὥστε at 740b35. In a first, weaker sense the text means:

[11] This passage should be read together with *GA* 2.1, 734b31: "We may allow that hardness and softness, stickiness and brittleness, and whatever other qualities are found in the parts that partake in life and soul, may be produced by heat and cold, but when we come to the reason (λόγος) in virtue of which flesh is flesh and bone is bone, that is no longer so. What produces them is the movement set up by the generator, who is in actuality what that out of which the offspring is made is in potentiality."

[12] Cf. *GA* 2.4, 740b36: "If it is the nutritive soul, this is also what generates the offspring (εἰ οὖν αὕτη ἐστὶν ἡ θρεπτικὴ ψυχή, αὕτη ἐστὶ καὶ ἡ γεννῶσα)."

the matter of the growth and the matter of the embryo being the same, the productive power of these two processes is presumably the same. But in a second, stronger sense we may understand that an identical matter cannot be acted upon but by an identical power.

That growth and preservation are the same is explained in texts like the following one:

> The essence of nutriment and the essence of the factor producing growth are different. For it is in so far as the ensouled being has a certain quantity that <nutriment> is capable of producing growth, but in so far as it is an individual substance, it is nutriment. For the living being maintains its substance as long as it is nourished. <The nutriment> brings about generation, not of the being which is fed, but of something like it. For its substance already exists, and nothing generates itself, but rather maintains itself. So this principle of the soul is a potentiality capable of maintaining its possessor, while nutriment makes this principle capable of action. For this reason, if deprived of nutriment, the living being cannot exist. (*DA* 2.4, 416b11–20)[13]

At this point, the progress made is not a small one. Generation no longer depends on the nutritive soul only because the semen and the menses are the final products of a digestive process. As the passage from *GA* 2.4 quoted above suggests, the nutritive soul makes the animal grow as it also "in the same way constitutes <the embryo> from the beginning." The nutritive soul does so "in the same way," i.e., in using the hot and the cold as its tools to introduce a determinate ratio in the matter out of which the animal is made. This means that nutrition is not merely accumulation of nutriment in the stomach, because to turn this food into flesh and bones the nutritive soul has to introduce a ratio in it, viz. a proportion between the elements in the nutriment. In the same way, the nutritive soul has to introduce a ratio in the matter found in the womb of the female to turn this matter into an embryo. In other words, the nutritive soul is the faculty that has the power to introduce the due proportion between the components of the nutriment to make it flesh and bone or to make it an embryo.[14] At this point it should be clear that the digestion for the preservation of the

[13] Cf. *GC* 1.6, 322a25–28: "Nourishment is the same as growth, but its being is different. For as the body which accedes is potentially a quantity of flesh, it makes flesh grow, but as it is potentially flesh only, it is nutriment."

[14] There is a difficulty I want to mention here. During the generation of an animal (in the case of an animal reproducing by the intercourse of a male and a female), which nutritive soul is acting to introduce the adequate ratio in the matter in the womb of the female? Is it the nutritive soul of the male or of the female? Passages like *GA* 2.4, 740b24 seem to suggest that it is the female soul. But in *GA* 2.1, 734b31, Aristotle explicitly says that this is the role of the male.

animal, the digestion for the growth of the animal (which takes place only in a due period of the life of the animal, whereas plants grow continuously[15]), and the formation of the embryo are really one and the same process, even if their *being* is different. The distinction of things that are one and the same thing, but of which the *being* is different is usual in Aristotle.[16]

Consequently, the *GA* is also a treatise on nutrition – not only because it deals with the generative residues elaborated from nutriment (semen and menses), but also because the process of formation of the embryo is the realization of one of the functions of the nutritive soul. From the point of view of this stronger reason for the identity between nutrition and generation, the theory of generation is in fact a special case of the theory of nutrition, and not the other way around, even if (as the passage *On the Soul* 2.4, 416b23, quoted earlier says) the supreme goal of a living being is to reproduce itself. In other words, according to the weaker reason, the theory of the elaboration of nutriment into semen is an aspect of the theory of the generation of the animals, whereas according to the stronger reason the theory of generation is in fact a chapter in the study of the various functions of the nutritive soul, since the generative motion (which ends in the formation of the embryo) is a particular case of the motion through which the nutritive soul elaborates nutriment according to a definite ratio (λόγος).

At this point, we are facing another priority problem, which is quite enlightening. As I noted, there is a treatise in the Aristotelian corpus that deals precisely with our three vital operations – γένεσις, αὔξησις, and ἀλλοίωσις – in searching for their common principles. This is the treatise *On Generation and Corruption*. But, in this work, alteration and growth (along with mixture, which is not under discussion here) are, as Marwan Rashed says, "weak generations."[17] This means that generation is a more radical transformation than the other vital transformations. And this seems to be quite obvious and easy to understand, as γένεσις is a change according to substance, whereas αὔξησις and ἀλλοίωσις are changes in the categories of quantity and quality, respectively.

Now, in *GA* 2.4, 740b35, we are told that the power by which what is generated in the first place is produced is stronger than the power that

[15] Cf. *Long.* 6, 467a12: "Plants continuously renew themselves and hence live a long time. For new shoots continuously come and the others grow old, and the same happens with the roots."

[16] E.g. *Phys.* 5.5, 229a29–30: "Recovering health and leaving illness are two identical changes, but their essence is not the same."

[17] Rashed 2005: LXXXV.

produces growth. It seems that in this case "growth" here encompasses two functions: the maintenance of the body and what is, strictly speaking, growth. That the generative power is greater than the power that produces growth certainly means that a greater effort is required to constitute the embryo than to make it grow. That said, we may assume that, once semen and menses have been produced, the motions required to form the embryo are weaker than those required to make blood out of cauliflower. One can say "weaker" precisely because these motions share the same nature: a new λόγος, which is the definition of the final state, must be introduced into the material components of the previous state. This new λόγος is produced by the action of the hot and the cold. Now, the difference between the λόγος required to make the cauliflower out of the three or four elements it is composed of and the λόγος required to make blood is greater than the difference between the λόγος that is required to make blood and the λόγος that is required to make the embryo.

How are we to reconcile these two conflicting views about what is prior to what? Here we have to invoke another kind of priority relation. This time we are not dealing with what is prior because it is a goal vs. what is prior because it is more basic; rather, we are dealing with the distinction between what is prior because it is more radical vs. what is prior because it requires more strength. That, on the one hand, the elaboration of nutriment into blood requires more energy that the transformation of blood into semen and of the semen into an embryo was probably not hard to accept in Aristotle's time. The second process, in effect, takes place within the living being itself, contrary to the first one, which has to transform something external to the animal into an organic part of the same animal. Still, it would be better to conceive of growth (or maintenance of the body) as a kind of generation of oneself, rather than to conceive of generation as a kind of growth. Even in the uterus, one cannot say that the embryo is the result of a process of growth, although it is made out of the preexisting female matter. The embryo is *generated*, which means that Aristotle is an epigeneticist and not a preformationist.[18] Therefore, generation is the most radical and the most complete of the natural motions which take place within an animal.

The *GA* is a treatise on the nutritive soul in the sense that it considers the elaboration of nutriment by the nutritive soul by the means of the hot and the cold, even if it considers this process mainly in its more complex and radical form, i.e., as the generation of a living being. At the level of

[18] I refer the reader to Chapter 5 in this volume. Cf. Kullmann 1979.

generation, this elaboration is a combination of two related but different processes, i.e., the elaboration of blood (which is elaborated food) into semen and menses, and the elaboration of female matter (which is itself elaborated blood) into an embryo by the motions internal to the semen.

Some Problems with Aristotle's Theory

Such a unified doctrine, which considers maintenance of the body, growth, and generation as similar processes beyond their differences, is problematic in several ways. Let me focus on a couple of problems.

The first concerns the role of the female. Aristotle explains why a female cannot achieve generation by herself (*GA* 1.21, 730a28). It is usually said that this happens because, being too cold, the female is unable to produce the motions which are supposed to inform the menses. But this is not exactly what Aristotle says. In fact, according to Aristotle, the female cannot generate by herself because the sensitive soul is transmitted by the male. Aristotle even says that the sensitive soul is "the essence of the male" (*GA* 2.5, 741a15). The female is therefore unable to produce the sensitive soul. She possesses a sensitive soul (since otherwise she would not be an animal), which she got from her father. But what about the nutritive soul? Is it also transmitted to the embryo by the male, as it is generally understood? Given what I said earlier, it seems that the female does not need a male to transmit the nutritive soul to her embryo. Of course, in a superior animal like the human being, we are not facing this problem, because a woman gets pregnant only after receiving the semen in her uterus. But the case of wind-eggs, to which Aristotle refers just after he has explained why females cannot generate by themselves, seems to point in the direction of the transmission of the nutritive soul by the female. Aristotle says that, since wind eggs can be corrupted, they possess life "in some way." He goes on to say that "it is therefore obvious that they have some soul potentially. But what kind of soul? It is necessary that it be the last one, i.e. the nutritive" (*GA* 2.5, 741a23).

A question Aristotle does not formulate but which is the counterpart of the one we have just considered is this: Why are males unable to generate by themselves? They possess the semen that contains the sensitive soul, and they also have the matter elaborated from the nutriment, namely blood. What they do not have is the appropriate matter to form the embryo, a kind of matter that, in one of the rare puns of my career, I called "diet sperm." For this female matter is just nutriment less completely elaborated than semen, because of the relative coldness of females. But if the male can produce a more elaborated state of the nutriment, namely semen, he should be able to produce a less elaborated version of it. At the end of the day, Aristotle offers no convincing physiological reason for the incapacity of males to generate by themselves.

Conclusion

Let us come back to what we started with. What makes a living being what it is is its capacity to assimilate its environment. This capacity comes from three activities that make the living being able to survive, grow, and reproduce. In the case of an animal, its growth stops at one point, contrary to what happens in plants, and its generation presupposes that it be endowed with a sensitive soul during its formation as an embryo, since animal is defined by sensibility. If Aristotle gathered nutrition and reproduction in one function, it is probably because he ultimately conceives of the reproduction of living beings, and especially of animals, as a sort of self-cloning. Out of the very motion by which the animal constitutes itself, it reproduces itself. As he is an epigeneticist, Aristotle of course does not think that there are small "animalcules" already formed in the female uterus. He does think that the embryo undergoes a process of formation. But generation is just the actualization of the capacity to introduce order and organization into a matter, this capacity being that of the nutritive soul. The great gap to be closed is that between the nonliving and the living. This happens when nutriment is turned into blood. After this, there is natural internal development of the living.

CHAPTER 5

Aristotle on Epigenesis
Two Senses of Epigenesis

Devin Henry

Introduction

In a recent paper, Aryeh Kosman draws attention to the fact that Aristotle rejected the ancient theory of preformationism in favor of his own model of epigenesis:

> Aristotle is clear and forceful in his rejection of the notion that the offspring itself must somehow be from the very beginning in the seed of one or the other of its parents (*GA* I.21.729b1ff.). His rejection is made possible by a theory that employs the important concept of power or potentiality – δύναμις – according to which the animal is contained in the seed only in the sense that the seed has the power to produce the animal by a process of what came to be called epigenesis. This paper is primarily a plea for the recognition that Aristotle's theory is through and through a theory of epigenesis, and an argument about what follows from that fact.[1]

Of course, it should come as no surprise that Aristotle was an epigenesist in the sense that he thought form and structure emerged gradually from an unorganized, amorphous embryo. This much, at least, has become somewhat of a platitude. Perhaps the clearest endorsement of Aristotle's commitment to epigenesis comes at *GA* 2.3, 736a35–b5. There he tells us that the animal's form (soul) isn't acquired *in toto* at the start of development; rather, it comes into being gradually as the process advances. While the early embryo already possesses certain rudimentary nutritive capacities, it isn't until later that it comes to possess the capacity for sensation (the property that makes it an animal), while those features that make it a member of a particular species are not acquired until the end of the

I am grateful to Alan Love whose correspondence on epigenesis provided the initial catalyst for this paper. I would also like to thank Mariska Leunissen for her engaging comments on an earlier draft of this paper.

[1] Kosman 2010: 149–50. See also Peck 1942: 144.

process. That is why Aristotle says in that passage that a developing organism does not come to be an *animal* and a *human*, or an *animal* and a *horse*, at the same time; rather, the generic attributes are acquired before the specific ones. This expresses, rather clearly, a commitment to epigenesis. For if preformationism were true, then it would be the case that the fetus comes to be an animal and a human, or an animal and a horse, at the same time; each one would be a human or a horse *from the very beginning*. And since Aristotle thinks that capacities of soul develop together with the parts that execute those functions (e.g., sight develops along with the eye), to say that the parts of the soul emerge gradually also entails that the parts of the body do so as well (for "there is no part of the soul which is not in some part of the body," *GA* 2.1, 734a15). So neither form nor structure preexists in the embryo except in potentiality.[2]

But there is a more interesting question that Kosman fails to consider. Modern developmental biology now recognizes two senses of "epigenesis" that give rise to what Alan Love characterizes as "a pervasive ambiguity lumbering around in the literature."[3] In the first place, "epigenesis" is used for that familiar idea concerning the gradual emergence of form and structure (the sense of epigenesis that Peck and Kosman have in mind). It is opposed to the view that the complete organism preexists in miniaturized form inside the sperm (or egg), so that what we call "development" is really just the augmentation of those preformed structures. But biologists also use "epigenesis" to emphasize the context-dependency of the process itself.[4] According to this version, developmental pathways are not fixed ahead of time by the genetic program. Rather, the genes represent a set of potential pathways. And which pathways are actualized during development is determined in real time as the process unfolds in response to environmental cues (including factors internal to the developing system but external to the genome).[5] On this view, then, all the information to build a new individual is not already contained in the genome from the start. Instead, a developing embryo is viewed as a more dynamic and responsive system that reacts to real-time inputs from the internal and external environment. I shall call these two senses of epigenesis "Epigenesis-1" and "Epigenesis-2," respectively:

[2] See also: *GA* 2.4, 739b34–740a4; 2.5, 741b7–15. [3] Love (personal correspondence).

[4] Müller and Olsson 2003: 114.

[5] Epigenesis in this sense is sometimes characterized as environmentally induced phenotypic variation (Müller and Olsson 2003: 117).

EPIGENESIS-1: The complete organism does not preexist in the seed from the start; rather, form and structure emerge gradually from an unorganized, amorphous seed.

EPIGENESIS-2: Development is not simply the unfolding of a predetermined sequence of changes specified in advance by the organism's genotype. It is also sensitive to inputs from the internal and external environment, which help determine *in real time* which of the many potential pathways are actualized during the process.[6]

While it is certainly not news to anyone that Aristotle rejected preformationism in favor of Epigenesis-1, it is completely up for grabs whether or not he thought development was "epigenetic" in the sense of Epigenesis-2. This is important because Epigenesis-1 is completely compatible with a rejection of Epigenesis-2. Someone can hold that form and structure emerged gradually from an unorganized, amorphous embryo and yet still think that the process unfolds according to a fixed sequence of preprogrammed changes. My hunch is that this was in fact Aristotle's view.

There is a swift argument and a more complicated argument for my reading. The swift argument is as follows. In *GA* 2.1 Aristotle compares the process of development to the gradual weaving of a net, an image he borrows from the poems of Orpheus (734a18–20). This clearly points toward Epigenesis-1. But he also goes on to drawn an analogy with self-moving automatons:

> It is possible for *A* to move *B*, and *B* to move *C*, and for it to be like the self-moving automatons. For while at rest the parts of the automaton somehow contain [the movements] in them *in capacity*, so that when some primary external agent sets the first part in motion, straightaway (εὐθύς) the next part comes to be [in motion] *in activity*. (*GA* 2.1, 734b9–13)[7]

> As the parts [of the animal] are already present in the matter *in capacity*, once the starting-point of motion is acquired they develop in a connected sequence just like the self-moving automatons. And the meaning of the statement made by some natural scientists about "like making its way to like" must not be understood as saying that the parts of the body move in the sense of changing their location [. . .] but in the sense of coming to be *in activity* what they previously were only *in capacity*. (*GA* 2.5, 741b7–15)

[6] Of course there is much more to Epigenesis-2 than just this (see Müller and Olsson 2003). In this chapter, I shall focus exclusively on adaptive developmental plasticity, the ability of an embryo to adjust its phenotype "on the fly" in response to changes in the (internal and external) environmental conditions.

[7] All translations are my own unless otherwise indicated.

This points us toward a rejection of Epigenesis-2. By saying all the parts of the animal are already contained in the embryo δυνάμει, I take it Aristotle means that an embryo already has the capacity to develop those parts,[8] just as the automaton already has the capacity to execute a fixed sequence of movements while at rest. Once the process has been triggered by the external agent, the whole pattern of motions unfolds in a connected sequence. Aristotle's choice of model here suggests that he envisions development as a prearranged sequence of transformations. For the automatons he has in mind are not dynamic AI systems capable of modifying their behavior on the fly in response to environmental cues but deterministic mechanisms whose movements are fixed by their original construction.[9] The mechanical automaton thus points us toward a model of development where all of the changes are specified in advance by what Allan Gotthelf calls the animal's "internal potential for form."[10]

The more complicated argument for my view, which I develop in the remainder of this paper, draws on Aristotle's theory of causal powers from *Metaphysics* 9 (*Theta*). Given Aristotle's views about the different kinds of causal powers, it looks like he thinks only intentional agents endowed with actual decision-making powers could be capable of the sort of plasticity at the core of Epigenesis-2. Epigenesis-2 seems to involve powers for alternative outcomes, and Aristotle is explicit that such powers require something like a rational decision (προαίρεσις) that controls which of those alternatives to bring about. If I am right, then he could not have made sense of the idea of a developing embryo (as a nonintentional system) making adjustments to its phenotype on the fly in response to emerging problems and opportunities, given the conceptual resources available to him.[11]

[8] As Kosman says: "the animal is contained in the seed only in the sense that the seed has the power [δύναμις] to produce the animal" (quoted earlier in this chapter).

[9] See *MA* 7, 701b1–10. I discuss Aristotle's use of automatons as a model for development in Henry 2005.

[10] Gotthelf 2012: 3–44.

[11] This way of putting the point might be too strong, since Aristotle would have likely allowed that animal behavior displays a certain amount of adaptive plasticity (e.g., a lion might alter its course on the fly while pursuing a gazelle to avoid unexpected obstacles) even though animals lack "decision" in the technical sense of προαίρεσις. In this case I imagine he would ascribe the plasticity of behavior to the possession of various perceptual capacities, including imagination (φαντασία) and certain higher-order discriminatory powers (τὸ κριτικόν) that are at least analogous to rationality. See *DA* 3.2, 426b8–472a15 (cf. 432a15–17) and the chapters on ethology in *HA* 8. This possibility does not affect the argument of this paper since it is unlikely that Aristotle thinks the behavior of a developing animal owes anything to such complex perceptual capabilities. I am grateful to Fermin Fulda for pressing me on this point.

Aristotle and Epigenesis-2

According to Aristotle's theory of heredity the father's semen contains a set of "movements" (κινήσεις) drawn from a corresponding set of "potentials" (ἀπὸ τῶν δυνάμεων), each of which is a capacity for developing a different part of his body (*GA* 4.3, 767b23–26, 768b1–5).[12] On this model an embryo has, from the very beginning, a set of developmental capacities in virtue of which it is capable of developing a certain shape and form. Following Gotthelf, I shall refer to these capacities collectively as the embryo's potential for form (where "form" is used broadly for what we would call the animal's phenotype). Our question is about how much of the animal's phenotype Aristotle thinks is specified in advance by this inherited potential for form. As a working example consider the account of spurs and talons from *PA* 4.12 (694a9–28). Aristotle tells us that some of the heavy birds have spurs on their legs for protection to make up for the fact that they do not fly well, while raptors (which are powerful fliers) instead develop talons on their feet. But no bird has both: spurs are unnecessary if you have talons, and talons are an impediment for ground birds since they would get stuck while walking. Aristotle then says:

> It is from necessity that this difference comes about during development. For the earthen effluence in the body becomes useful for aggression: when it flows upwards, it makes either hard beaks or large beaks, while if it flows downwards it makes spurs on the legs or large and strong claws on the feet. But it does not make each of these in different places simultaneously; for were it spread about, the nature of this residue would become weak. (*PA* 4.12, 694a23–8, Lennox translation)

By saying that the difference comes about "during development" one could take Aristotle to mean (in accordance with Epigenesis-2) that the potential for form inherited by each type of bird is undetermined with respect to the parts in question, so that all that is specified in advance are *potential* developmental pathways. On this reading it is left open at the start of the process which type of part a particular bird will develop. Instead, how the raw materials get used is not determined by the embryo's formal nature until the appropriate stage of development.[13] The challenge for this interpretation is to show that Aristotle actually has the conceptual resources to account for this kind of adaptive developmental plasticity

[12] Henry 2006b and 2009.

[13] The most sustained defense of this interpretation in the literature is offered in Leunissen 2013. I consider her view in the section "Nature as Craftsman" later in this chapter.

where embryos are able to adjust their phenotypes "on the fly" in response to novel problems and opportunities that emerge during the course of development. My suspicion is that he does not.

As a way into this issue consider how Aristotle conceptualizes development in *GA* 2.4:

> Concerning the differentiation (διάκρισις) involved in the formation of the parts this does not occur, as some say, through the natural action of like making its way to like [. . .] but rather owing to the fact that the residue of the female is the sort of thing that is potentially like what the animal is by nature, that is, that the parts are present in it potentially, though none of them are there in actuality. Each of them comes to be on account of this cause and also owing to the fact that when agent and patient interact (θίγωσιν) with one another in the way in which the one is active and the other passive [. . .] then *straightaway* the one acts and the other is acted upon (εὐθὺς τὸ μὲν ποιεῖ τὸ δὲ πάσχει). (*GA* 2.4, 740b12–24)

This passage gives us a glimpse into the basic causal model that Aristotle is working with in the *GA*. According to this model the development of a given part is conceptualized, in accordance with the definition of change from *Phys.* 3.1–2, as the actualization or fulfilment of the matter insofar as it is potentially that part. And that change is occasioned by the mutual activation of two correlative powers: (1) a passive power to be transformed into the part, which is rooted in the embryo's matter (cf. *GA* 2.5, 741b7: "the parts are present in the matter in potentiality") and (2) an active power to transform the matter into that part, which is a capacity of the formal nature (or generative soul).[14] Now one way to cash out Epigenesis-2 in Aristotelian terms would be to say that the inherited potential for form that governs development includes certain active powers for producing alternative effects depending on the developmental context.[15] So, for example, Aristotle might hold that the potential for form inherited by raptors and fowl includes a capacity to develop spurs *or* talons. Which way that capacity is activated is not specified in advance; rather, it is determined on the fly in response to the teleological demands that emerge from its particular way of life (*PA* 4.12, 694a9–22).

This last part is something we might expect in the context of Aristotle's theory. Recall that Aristotle thinks animals develop their generic form before their specific form, so that the embryo does not come to be a bird

[14] Aristotle associate the active powers of development with the embryo's generative soul – which he identifies with the *nature* of each thing – just after our passage at *GA* 2.4, 740b34–741a3.

[15] Cf. Leunissen 2013: 19.

and a raptor or a bird and a fowl at the same time (*GA* 2.3, 736a35–b5). Rather, the features that make the embryo a raptor or a fowl are not acquired until the later stages of development. This means the developing bird will not actually be subject to any particular design requirements stemming from any specific way of life until those later stage. So it would make sense for Aristotle to say that whether the particular bird develops spurs or talons isn't determined until later in the process of development, since there is nothing that teleologically necessitates one part or the other until the later stages. (Neither spurs nor talons are *conditionally necessary* at the start.) If this is right, then we might expect Aristotle to say that the embryo's formal nature cannot make the "decision" to produce spurs or talons until those later stages.

The trouble with this reading is that, while Aristotle does recognize capacities for bringing about alternative effects, he argues that such powers are always found "together with reason" (μετὰ λόγου).

According to *Meta.* 9 causal powers come in two kinds: rational and nonrational powers (9.2, 1046a36–b2). Nonrational powers are characterized by two essential features. First, each one is productive of a single effect only with no alternative, e.g., heat is only capable of heating and what is cold can only cool (9.2, 1046b6–7; 9.5,1048a8). Second, the activation of a nonrational power is necessitated by contact with its correlative power under the appropriate causal conditions so that, as soon as these conditions obtain, the power is activated straightaway (εὐθύς) if nothing impedes it (*Meta.* 9.5, 1048a5–8; compare the *GA* 2.4 passage from earlier in this chapter). By contrast, rational powers are powers to produce alternative effects, e.g., the art of medicine is productive of health *or* disease (*Meta.* 9.2, 1046a36ff.). For this reason, Aristotle says, contact with the corresponding power under the right conditions is necessary but not sufficient to activate it. Instead, there must be some further controlling factor (τὸ κύριον) that determines which of the two alternative ways of activating the power occurs, which he identifies as decision (προαίρεσις). For example, the medical art on its own cannot effect any change. The doctor must decide, on the basis of deliberation, whether she wants to heal her patient or make him sicker. Aristotle makes this point in *Meta.* 9.5:

> Some things are capable of effecting change in accordance with reason and their powers are accompanied by reason, while others lack reason and their powers are non-rational, and it is necessary for the former to be present in ensouled things while the latter are present in both. And with powers of the latter sort, when agent and patient interact with one another in the way in which they are capable, it is necessary (ἀνάγκη)

> that the one acts and the other is acted upon, while in the former case this is not necessary. For, while all non-rational powers are productive of a single effect only, rational powers are productive of opposites so that one power will simultaneously produce contraries. But this is impossible. Therefore, there must be some other thing that exerts control (τὸ κύριον), namely, desire or decision.[16] For whichever outcome it desires decisively (κυρίως), it will produce this when it is in those circumstances that are appropriate to its capacity and associates with the patient. (*Meta.* 9.5, 1048a2–13; cf. *DA* 3.9, 433a1–6)

Given this conceptual scheme it is hard to see how Aristotle could make sense of Epigenesis-2. For the plasticity involved would seem to require the embryo to make literal decisions about which of a series of alternative pathways to actualize during its development.

Consider the raptor again. On the version of the reading developed earlier in this chapter, what a raptor inherits from its parents is a flexible δύναμις for producing talons *or* spurs. Which way that δύναμις gets activated is determined on the fly at the appropriate stage of development in response to certain environmental cues. But we have just seen that a δύναμις for bringing about alternative effects requires some additional factor to control its activation. And on Aristotle's theory this role is played by decision, which is a rational desire to bring about what deliberation has shown to be the best course of action. In that case it looks like Aristotle would have to say that the embryo (or its formal nature) literally decides that it would be better to produce talons rather than spurs given the demands imposed on it by its emerging way of life. For his causal theory doesn't countenance any other (nonrational) controlling factors for determining which way to activate that δύναμις. This gives us a reason to be skeptical of the idea that Aristotle accepted Epigenesis-2.

If all of the powers at work in development are of the nonrational sort (as Aristotle clearly thinks they are), then this leaves little room for the sort of adaptive developmental plasticity that is at the core of Epigenesis-2. First, as we have seen, every nonrational power is productive of a single effect only with no alternative. Second, the interaction with its correlative power necessitates its activation ("when agent and patient interact with one another in the way in which they are capable, it is necessary that the one acts and the other is acted upon"). In that case there is no need for any

[16] These amount to the same thing, since "decision" is defined as a rational desire to bring about what deliberation has shown to be the best course of action (*NE* 3.3). Decision in this sense is the efficient cause of action (in the sense of πρᾶξις), which Aristotle identifies as a single state with motivational and intellectual dimensions (*NE* 6.2).

additional controlling factor; the activation of each power follows immediately from the causal conditions ("when agent and patient interact with one another in the way in which the one is active and the other passive . . . then straightaway the one acts and the other is acted upon"). If this is right, then the actualization of the inherited potential for form that constitutes development should unfold in a cascading series of automatic changes triggered by the interactions of the various causal powers involved under the appropriate conditions. And that is just what the model of a self-moving automaton leads us to expect.[17]

To close this chapter, I want to look at a recent interpretation of Aristotle's theory that attributes to him exactly the sort of view that I have argued he could not have entertained.

Nature as Craftsman

The most extensive defense of the epigenetic reading of Aristotle in the literature is Mariska Leunissen's paper, "Crafting Natures: Aristotle on Animal Designs."[18] Leunissen argues that Aristotle's tendency to depict nature as a tinkering craftsman throughout the biological works is a clear indication of his commitment to what I am calling Epigensis-2:

> At the physiological level, that is, when we translate Aristotle's talk of goal-directed natures into realizations of potentials for form in the manner proposed by Gotthelf . . . this means that those potentials have to be equally "creative" or "dynamic." If I am right in assuming that Aristotle's depictions of natures as craftsmen are no mere metaphors, but in fact reflect different causal patterns that underlie animal generation, the potentials for form that guide embryology and the later development of animals cannot be blind,

[17] By characterizing the actualization of the embryo's potential for form as a "cascade of automatic changes," I do not mean to suggest that it is a purely mechanical process. A more complete explanation of the process would locate this (efficient causal) account within a broader teleological framework by showing how each of these necessary interactions as well as their overall choreography take place for the sake of the adult form that results from them (*GA* 2.6, 735a36–b8). Thus I agree with Leunissen that the capacities at work in development are not "blind"; they are goal-directed capacities. But it does not follow that they cannot also be "automated self-regulating principles." (How a process can be necessitated and also for the sake of its end is outside the scope of this paper: see Johnson 2005, chapter 7.) The question of whether development is "blind" or "for the sake of its end" is a separate issue from the question of whether it unfolds according to a predetermined pattern or whether some changes are determined on the fly in response to environmental cues.

[18] The main impetus for this chapter came out of conversations with Mariska Leunissen on her original draft. It is worth pointing out that I was the one who originally encouraged her to characterize her position in strongly epigenetic terms, as I thought it best captured the main thrust of what she was trying to say.

> automated self-regulating principles.[19] Rather, they are flexible and complex, and include the capacities to make the best use of extra materials and to respond to emerging problems or possibilities (Leunissen 2013: 19).
>
> This indicates that at least *at the level of embryogenesis*, Aristotle does allow for something like a transformation of species and that his view of development is epigenetic in the true sense of the term. That is, the "guidelines" for building with which nature works do not predetermine in advance all the changes that need to be made, but rather leave room for nature to make adjustments to animal design "on the fly," as the development proceeds. For his theory of natural teleology this means that the underlying physiology must be rather complex and flexible: the potentials for form that guide the realization of those species are dynamic, rather than blind, automated self-regulating principles, as they are shown to be capable of responding to "unexpected" changes in circumstances (Leunissen 2013: 29, emphasis in original).
>
> My reconstruction above of Aristotle's assumptions about how natures "weigh" the available natural possibilities and choose the best design option for each individual kind of animal shows that many features are not "preprogrammed" by the definition of the substantial beings of animals. The "phenotype" of animals develops gradually, as a result of natures not just doing what is necessary, but also doing what is better and best in response to emerging problems and opportunities (Leunissen 2013: 44).

On Leunissen's reading, then, the potentials for form passed on in reproduction do not specify in advance all of the natural changes that occur during development. Rather, they leave room for nature to make adjustments to the animal's basic design on the fly as development unfolds. As should be clear by now I am not convinced that this picture accurately describes Aristotle's view.

In the first place none of the examples Leunissen cites in favor of her interpretation commit Aristotle to Epigenesis-2. In each case Aristotle talks about nature making "modifications" and "adjustments" to a kind's basic body plan in order to optimize the form of some particular species. But nothing in his analysis suggests that he is thinking of these as real-time embryological responses to novel design problems that emerge only during the process of development. Consider the example of the elephant's trunk (Leunissen 2013: 34). Normally four-footed animals with split toes have feet that provide support for their bodies and a means for transporting food to their mouths. However, given the constraints on the elephant's form,

[19] As mentioned "automated" does not entail "blind" (see note 17).

Aristotle says that nature was forced to modify this basic design by doubling up functions on the trunk (*PA* 2.16, 659a20–30, 34–6). Leunissen claims that in this case the design problems faced by the elephant emerge during its formation, so that nature has to modify its basic body plan as the embryo develops: "In cases like this, nature still realizes all the functions specified by the definition of the substantial being of the animal, but, because of the design-problems that only become clear during the animal's production process, it has to do so in an 'unexpected' way" (Leunissen 2013: 34–5). But there is no reason to think that. While Aristotle clearly thinks nature "tinkers" with her designs in some sense, nothing he says in that context suggests that he is thinking of these cases in terms of nature making real-time embryological adjustments to the developmental system. In each case he could simply be presenting us with a thought experiment where nature makes certain hypothetical adjustments to the species original design (like an engineer tinkering with the design of his automaton before setting it in motion).[20] In other words, all of Leunissen's examples are compatible with thinking of the modifications in question as being programmed into the inherited potential for form from the start. Leunissen at least provides no reason to think her own epigenetic reading of those examples is to be preferred.

Leunissen's strongest case comes from her analysis of *PA* 3.2, 663b22–35 (Leunissen 2013: 35–6). There Aristotle tells us that certain surplus "residues" can sometimes get co-opted by nature for some additional purpose during the process of development:

> But, since there is a necessary nature, we must say how the formal nature makes use of things present of necessity for the sake of something. [. . .] For the surplus residue of this kind of [sc. bony] body, being present in the larger of the animals, is used by nature for protection and advantage, and the surplus, which flows of necessity to the upper region, in some cases it distributes to teeth and tusks, in other cases to horns. (Lennox translation, with modification)

According to Leunissen, these leftover materials are not present because they are necessary prerequisites for the construction of parts that are themselves specified in the definition of the animal's substantial being

[20] Leunissen (2013: 31) herself recognizes several examples where Aristotle is speaking about some "hypothetical moment in time" where nature first designed an animal, analogous to the "first creation" of human beings in the *Timaeus*. I see no reason to think that this example is any different. This moment is necessarily hypothetical, since there is no time for Aristotle before elephants existed since species forms are (in some sense) eternal.

(they are not conditionally necessary). Instead they are accidental (unforeseen) by-products of other processes that operate through material necessity. Once those materials become available, the nature of the embryo immediately co-opts them to make what Leunissen calls "bonus features," parts the animal could have done without but make it better. Leunissen takes this as evidence of the adaptive plasticity of development; for these surplus materials are accidental and not specified in the original plan for constructing the animal.

But Aristotle need not be thinking of the availability of surplus residue as an unforeseen accident that nature happens to exploit opportunistically. The text is perfectly compatible with thinking that the use of these materials is something that is also specified in advance by the developmental program. Imagine an expert sculptor who has been making the same statue for the entire length of her career. She knows that she always needs two-thirds of a cubit of clay for the hands. But she also knows that her supplier packages clay in whole cubits, so that it normally turns out that she has one-third of a cubit left over; the surplus is a necessary by-product of how the clay is packaged. Since she is not wasteful, she always plans on that surplus to make the wings on the helmet. In this case the surplus materials are not some unforeseen accident that affords her an opportunity, which she decides *in the moment* to use for the wings. It is something that happens either always or for the most part, and so she always anticipates having some left over clay to make those extra parts. It is something she budgets for in her original design. Aristotle could have something similar in mind at *PA* 3.2, 663b22–35.

The other worry with Leunissen's reading is the one already mentioned. Given the conceptual resources available to Aristotle, it is difficult to imagine how formal natures could be responsible for making real-time adjustments to the developmental system in the way Leunissen suggests without turning them into rational agents. This gives us a reason to reject her epigenetic interpretation, since Aristotle does not think formal natures make literal decisions about how to realize their ends.

On Leunissen's epigenetic reading Aristotle treats the development of an animal as the actualization of an inherited potential for form that guides the actions of the formal nature in constructing the animal in the way that the blueprints guide the actions of the builder. But (she argues) that potential for form does not completely determine its actions. Instead, just as a craftsman can make adjustments to the blueprint of the house on the fly in response to problems that arise during its construction, animal natures are capable of making real-time modifications to the kind's basic design during the course of its development. In these cases the inherited potential for form only specifies a set of potential pathways. And which of those pathways is actualized during development is "up to" the animal's formal nature. The key to her reading is the claim that the capacities that guide development are not "blind, automated self-regulating principles"; rather, "they are flexible and complex, and include the capacities to make the best use of extra materials and to respond to emerging problems or possibilities" (2013: 19).

We have already seen that this way of conceptualizing the capacities at work in development (as powers for alternative effects) requires the introduction of a rational agent whose decisions (rational desires for bringing about the conclusion of deliberation) control how those capacities get activated. For Aristotle does not countenance any other kind of nonrational controlling factor that could play that role. Consider what Leunissen says about snakes:

> A quick thought experiment reveals that no blooded animal whose length is out of proportion to the rest of their body would be able to move swiftly with either two or four feet, and in order to remedy that design problem, nature "decided" not to produce feet in such animals. With this information, it would then be possible to formulate a scientific explanation that does not appeal to any conscious intentionality in nature.

If the argument in this paper is correct, then it does not seem possible, on Leunissen's reading, to generate an explanation of this case without ascribing conscious intentionality to nature. Let's look at the example more closely.

Snakes are the only blooded land-dwellers that do not have legs. One reason for this, Aristotle says, is that nature does nothing in vain but always what is best for each substance given the range of possibilities (*IA* 8, 708a9–20). Now, since no blooded animal can move at more than four points of motion (*IA* 7), the most legs a snake could have would be four. In that case giving snakes legs would be pointless, since they could not move very well with only two or four (cf. *PA* 4.13, 696a12–15). So having no legs turns out to be optimal for a snake.[21] On Leunissen's reading the absence of legs in snakes is not something specified in advance.[22] Instead, all blooded land-dwellers inherit a "flexible" capacity for producing legs *or* their absence. During the development of a snake nature "decides" it would be better not to construct them and so makes the appropriate

[21] For my analysis of this case see Henry (2013, especially 232–3). [22] Leunissen (2013: 32).

adjustment by suppressing its activity. But how does this work exactly? Aristotle is clear that, when a capacity includes the potential for bringing about alternative outcomes, there must be some additional controlling factor – which on Leunissen's reading is the animal's formal nature – that makes a decision about which of those two alternatives to realize. By contrast, I have argued that if all the capacities involved in development are of the nonrational sort, then (*pace* Leunissen) there is no need for anything else to control their activation. The activation of each power is necessitated by contact with its corresponding power in the right causal conditions.

The Role of the Environment

In this chapter, I have argued that Aristotle simply lacked the conceptual resources to make sense of Epigenesis-2 without ascribing genuine rationality to nature. This is because he thinks only rational agents endowed with real decision-making abilities could be capable of the sort of plasticity at the core of that theory. At least Aristotle's theory of causal powers does not countenance any nonrational means for determining which of two alternative ways of activating a power occurs. Without this he simply has no way of explaining how a developing embryo (as a nonintentional system) could make real-time adjustments to its phenotype in response to emerging problems and opportunities. One might object here that this argument turns on the assumption that Aristotle's account of generation in the biological works is constrained by the theory of causal powers developed in *Meta.* 9.[23] If he is not working within that framework, then we are free to imagine other ways that the developmental capacities of embryos might be context-dependent without having to invoke rational decisions as the factor that controls which of several alternative ways of building the animal is realized. One possibility for this sort of nonrational controlling factor (τὸ κύριον) might be the environment itself.[24]

First of all, there is no indication that the biological treatises are not working within the framework of the *Metaphysics* theory of causal powers. On the contrary, the way Aristotle characterizes the formation of parts in the *GA* 2.4 passage ("when agent and patient come into contact with one

[23] Leunissen (personal correspondence).

[24] I am grateful to Mark Johnstone for raising this objection during a presentation I gave at McMaster University (February 2016).

another in the way in which the one is active and the other passive [...] then straightaway the one acts and the other is acted upon") is reminiscent of the description of nonrational powers in *Meta.* 9.5 ("with powers of the latter sort, when agent and patient interact with one another in the way in which they are capable, it is necessary (ἀνάγκη) that the one acts and the other is acted upon"). So I am not very sympathetic to this objection. But let's look at this alternative a bit closer. The question here is, in cases where the environment does make a difference to how things turn out, what is doing the explanatory work?

In Epigenesis-2 the capacity itself is the intrinsic (καθ' αὑτὰ) cause of the trait in question, while the environment is simply part of the causal conditions that determine how that capacity is realized (in environment A the δύναμις produces X, in environment B it produces Y). In this case the δύναμις itself is the efficient cause of the trait, while the environment forms part of the causal background that determines how it operates. Consider *PA* 4.12, 694a24–7 again. Aristotle tells us that some birds have hard beaks, some long beaks, some talons, and some spurs and that this difference comes about during development. Now suppose these four different parts are the intrinsic objects of a single δύναμις. According to the current suggestion, how that δύναμις is activated depends on certain features of the environment. For example, if the embryo develops in a hot and dry womb the δύναμις gets activated in one way causing the materials to flow upwards and form a hard beak. If the environment is cold and moist, that same δύναμις gets activated in another way causing it to flow downwards to form spurs on the legs. And so forth for the other options. In each case the "controlling factor" that determines how the δύναμις is actualized will not be a rational decision on the part of some intentional agent but the environmental conditions themselves.

This strikes me as a more promising way of defending the view that Aristotle endorsed Epigenesis-2. The place to look for such a reading would be *GA* 5, which is the most sustained discussion of the role of the environment in generation. Unfortunately the evidence that Aristotle thinks the environment plays this kind of causal role is weak. Instead he tends to treat the environment as a direct cause of features in its own right. A paradigm case is the formation of hair in wild animals:

> The condition of sheep in cold climates is opposite to that of man; the hair of the Scythians is soft but that of the Sauromatic sheep is hard.

> The cause (αἴτιον) of this is the same as it is in all wild animals: the cold air hardens and solidifies them by drying them; for as the heat is pressed out the moisture evaporates; and both hair and skin become earthy and hard. In wild animals, then, the exposure to the cold is the cause of hardness in the hair, while in the others it is their environment's being of such a kind. A proof of this is also what happens in the sea-urchins which are used as a remedy in stranguries. For these, too, though small themselves, have large and hard spines because the sea in which they live is cold on account of its depth (for they are found in sixty fathoms and even more). The spines are large because the growth of the body is diverted to them, since having little heat in them they do not concoct their nutriment and so have much residual matter and it is from this that spines, hairs, and such things are formed; they are hard and petrified owing to the congealing effect of the cold [sc. sea water]. (*GA* 5.3, 783a12–29, Platt translation with modifications)

In the case of the sea urchin, for example, the environment does not act as a trigger for the activation of a δύναμις for hard spikes. Instead the cold water itself is causally responsible for forming those parts: as the nutriment is streaming from its body during development, the cold congeals it into hard spikes. This is typical of the role Aristotle assigns to the environment in *GA* 5. But this is not what the epigenesist has in mind by environmentally induced phenotypic variation.[25]

[25] Sex determination might be seen as a counterexample to my argument, since Aristotle clearly thinks that whether or not an embryo becomes male or female is temperature dependent. (I am grateful to Maggie O'Brien for this point.) However, this is not epigenetic in the right way. Aristotle holds that "male" and "female" in the primary sense refer to the ability and inability to concoct blood into semen, respectively. And whether or not an animal acquires this dispositional property (and thus whether it becomes male or female) depends on how much the embryo is heated at conception (*GA* 4.1–2). So sex is influenced by temperature, but in a very direct way; it is a straightforward case of heating and cooling. The semen is hot and it attempts to heat the menstrual fluid to the same degree making it a male embryo. But sometimes the menstrual fluid is too cold in which case the semen fails to exercise its heating capacity sufficiently, resulting in a colder (female) embryo. There is nothing epigenetic about this account. Semen has an active power to heat the menstrual fluid, which it has in virtue of its own natural heat. And this is a standard nonrational power; it is a power for heating and only heating. If the matter is well-disposed to being heated, then the semen will exercise that power without impediment and assimilate the matter to its own hot nature resulting in a male embryo. If the matter is too cold relative to the heat in the semen, then the semen will not be able to exercise that power adequately leaving a cold (female) embryo. (See *GA* 4.1, 766a19–24; cf. *GA* 4.2 and Henry 2007, 255–7.) This is no more epigenetic than saying that fire has the power to heat water and assimilate it to its own temperature but that when the water is excessively cold it fails to heat it up to the same degree. Fire does not have a power for heating *or* cooling; it has a power for heating only. And that power can be successful or unsuccessful in bringing about its effect depending on the condition of the patient and/or the external environment.

Conclusion

At the end of her paper Leunissen (2013: 45) addresses the concern that her reading implies that Aristotle thinks formal natures deliberate and form conscious intentions in some literal way. She attempts to deflect this worry by pointing to Aristotle's cryptic remark toward the end of *Phys.* 2.8 that "even art does not deliberate" (199b26–30). Now it is certainly true that *artists* deliberate and make decisions about how to act based on their conclusions. But Leunissen argues that "the psychological states of the artist only matter in the sense that art cannot exercise itself, but operates through the artist" (Leunissen 2013: 45). But this is not quite right. It may be true that the τέχνη itself doesn't deliberate; after all, it is only a causal power. But *Meta.* 9.5 makes it clear that the decisions of the artist are not inconsequential. On the contrary, they are a necessary condition for the activation of that power insofar as those decisions control which of its effects comes about:

> For, while all non-rational powers are productive of a single effect only, rational powers are productive of opposites so that one power will simultaneously produce contraries. But this is impossible. Therefore, there must be some other thing that exerts control, namely, desire or decision. For whichever outcome it desires decisively, it will produce this when it is in those circumstances that are appropriate to its capacity and associates with the patient. (*Meta.* 9.5, 1048a8–13)

With rational powers (including crafts) the presence of the corresponding passive power under the right conditions is necessary *but not* sufficient to bring about the effect. There must be a rational desire on the part of the agent – a προαίρεσις – that controls which of the alternative ways of activating the power will occur. So the psychological states of the craftsman do matter a great deal. It is not just that crafts must be embodied to operate; without decisions crafts would be causally inert.[26] This is why I have argued that, if Leunissen's interpretation is correct, there most certainly would be a need to attribute genuine decisions to formal natures insofar as she makes the formal nature the controlling factor that determines which of the many alternative pathways specified by the inherited potential for form get actualized during development. On my reading, since all the capacities involved in development are of the nonrational sort,

[26] This strikes me as a persistent difficulty with Leunissen's work on natural teleology, namely, she never quite shows us how to get past the metaphor of nature as craftsman (see also Gotthelf and Leunissen 2012: 131–2).

there is no need for anything to make decisions about how to activate them. Their activation is necessitated by contact with their corresponding powers under the right causal conditions.

In this paper I have argued that, although Aristotle clearly defended Epigenesis-1 (a thesis about the gradual emergence of form and structure), there is no good evidence that he endorsed Epigenesis-2 (a claim about the ability of developing embryos to adjust their phenotypes in response to novel circumstances). Moreover, I have argued that the challenge for anyone who thinks he did endorse this kind of model is to show that he had the conceptual resources to make it work without assigning embryos real-time decision-making capacities. To end this paper let me point to one further reason for thinking that the whole notion of adaptive developmental complexity, so obvious to modern embryologists, was not even on Aristotle's radar.

In *Physics* 2.8 Aristotle argues that if there were no natures (understood as goal-directed principles of change), then embryogenesis would be completely random:

> Moreover, in the case of embryos (ἐν τοῖς σπέρμασι), anything must come to be at random. But the person who asserts this does away with nature and what exists by nature entirely. For those things are natural which, by a continuous process originating from an internal principle, arrive at some end. The same end is not reached from every principle, nor any chance end, but always the tendency in each is toward the same end, if there is no impediment. (*Phys.* 2.8, 194b14–18 translated after Hardie and Gaye)

It is the last sentence that is most relevant here: developing organisms have an intrinsic disposition (or nature) to move toward a certain end state, which they will achieve *if* there are no impediments. Aristotle often makes this qualification in connection with development. And one gets the distinct impression that he thinks the absence of impediments is a necessary condition for developmental success. What we never find (and this is the key point) is any suggestion that Aristotle thinks a developing embryo has the intrinsic capabilities necessary to adjust its trajectory in order to overcome impediments. This is striking. Aristotle devotes close to five chapters in *GA* to discussing the kinds of deformities that result when impediments derail the process. And yet there is no discussion (or even recognition) of the ability of nature to recover from such perturbations and avoid those kinds of monstrosities. If Aristotle really did recognize the possibility of Epigenesis-2, then we should expect at least one chapter devoted to that. For there is no doubt that Aristotle would have championed the capacity of embryos to "stay on target" in the face of impediments as the strongest possible evidence of nature's immanent teleological character. As Peter Calow (1976: 9) puts it, in these cases it's as if the developing embryo knows exactly what it wants to be and, even in the face of violent and unpredictable disturbances, still manages to achieve its ambition. But this is not something Aristotle seems to have noticed.

CHAPTER 6

A Latent Difficulty in Aristotle's Theory of Semen
The Homogeneous Nature of Semen and the Role of the Frothy Bubble

Marwan Rashed

Semen as a Homoeomery and the Weak *Aporia*

In his biological works, Aristotle sometimes uses the term "homoeomery" to describe semen (σπέρμα). It occurs, for example, in the canonical list in *HA* 1.1, 487a1–10.[1] In *HA* 3.2, in a lengthy and detailed listing of homoeomerous *parts* (511b1–10), semen and milk are not included, but in *HA* 3.20–22 Aristotle in fact provides quite long descriptions of both, thus (in the end) conforming to the list in *HA* 1. We cannot summarily sweep aside these passages by claiming that they reflect no more than a superficial and inadequate treatment of the notion; for in *GA* 1.18, 724b29, Aristotle sets it down in black and white that semen "is [. . .] a homoeomery (ὁμοιομερὲς [. . .] ἐστιν)." It is, however, undeniable that the *GA* passage does not claim to offer a definition of semen, but belongs rather to a discussion of its nature that is still in its preliminary stages.

When Aristotle does venture to offer a definition of a homoeomery, he bases it on the fact that each and every one of its parts is of exactly the same nature as the whole. Thus, according to *GC* 1.1, 314a20, "every [. . .] body in which the part is synonymous with the whole" is a homoeomery. In *PA* 2.9, 655b6, the relationship is presented the other way round. Homoeomeries are characterized by the fact that "the wholes <are> synonymous with their parts." In *PA* 2.2, 647b17, the text is no doubt corrupt,[2] but here once again we can detect the same idea expressed in

Translated from the French by Christopher Strachan.

[1] Also in *PA* 2.9, 655b23–27 and 4.10, 689a8–17.

[2] Christopher Strachan suggests to me the following correction. *PA* 2.2, 647b 17–20 reads as follows: καὶ γὰρ τῶν ὁμοιομερῶν ἡ διαίρεσις ἔχει διαφοράν· ἔστι γὰρ ὡς ἐνίων τὸ μέρος ὁμώνυμον τῷ ὅλῳ, οἷον φλεβὸς φλέψ, ἔστι δ᾽ ὡς οὐχ ὁμώνυμον, ἀλλὰ προσώπου πρόσωπον οὐδαμῶς. If, strictly speaking, homoeomeries are entities in which the part and the whole are named synonymously as defined in *GC* 1.1, then both part and whole share not just a name but also a definition. Where whole and part are named homonymously, as here, they share a name but are differently defined. As far as concerns their material composition such substances may be homoeomerous, but once the functionality of the whole is taken into account, they are treated by Aristotle as anhomoeomeries.

identical terms. When it comes to homoeomeries, "the part is homonymous with the whole."

These passages lead directly to a first *aporia*, which I shall call the "weak *aporia*." It raises the question of what it is that explains, in terms of matter, the size of embryos and their number, if semen is a homoeomery, and if a homoeomery is defined as a substance whose parts have the same name as the whole. When the seminal fluid comes into contact with the menses, there is nothing in the homogeneous indeterminacy of the homoeomery that appears to be able to explain either the number of embryos or their size.

It is not hard to find a way out of this difficulty. There are at least two means of countering it at our disposal. The first consists in distinguishing two senses in which Aristotle employs the term "homoeomery,"[3] or, to be more precise, in linking two possible denotations for one and the same sense of the term. The sense is the one just mentioned; the two possible denotations have to do with the biological function of the homoeomery in question. The homoeomery-1 will be the homoeomery considered in terms of its chemical formula, or, more exactly, as something that can only be envisaged as a chemical formula. Fundamentally what are involved are humors (which, because they are liquids, have no configuration of their own) and "soft" tissues which lack a distinct configuration. Rigid homoeomeries, on the other hand, most of the time, have a configuration dictated by an identifiable function. This characteristic brings them closer to anhomoeomerous bodies, of which Aristotle provides a clear description in *PA* 2.9, 655b15–23. Indeed, more than that, Aristotle appears to admit that one and the same body may be homoeomerous from one point of view and anhomoeomerous from another. This is what he says on the subject of the heart and the viscera in *PA* 2.1, 647a31–b9. Thus Aristotle goes a little further, even, than considering that certain homoeomeries may be *studied* along with anhomoeomeries by reason of their function: in a

Such is the case with the viscera described as τῇ μὲν ὁμοιομερῆ τῇ δ' ἀνομοιομερῆ at the end of *PA* 2.1 (647b8–9). In anhomoeomeries of this class, the part and whole are homonymously named, but their definitions are different. In our passage, it appears to be these that are distinguished from the (οὐχ ὁμώνυμον) class of anhomoeomeries in which individual parts consist of incomposite, homoeomerous substances such as flesh, bones, and sinews: *HA* 1.1, 486a7–8 makes clear that πρόσωπον belongs to this class (cf. *GC* 1.5, 322a19, and see also 321b 21–22). Though the parts of a complete vein are also called veins, the parts of a face are not called faces. It may be that (with the change to <ἀν>ομοιομερῶν) this passage belongs immediately after 647b8–9, rather than where it is in the middle of a discussion of homoeomeries of which, somewhat confusingly, φλέψ has figured as an example immediately before it.

[3] See Furth 1988: 80–2.

certain respect, the function is *in fact* the function that anhomoeomeries have. It might be added that Aristotle is at the same time well aware that it would be quite arbitrary to treat certain homoeomeries, including those with a manifestly functional configuration, in such a way as to isolate them too much from the organic system to which they belong, as he says in *PA* 2.9, 654a32–b13 on the subject of the "nature of bones and of blood-vessels." This passage well illustrates the limitations of basing a moriology on too strict an interpretation of the homoeomerous/anhomoeomerous distinction. Taken by itself in isolation from anything else, "part" is effectively devoid of meaning: for its function to be properly understood, it needs to be looked at in conjunction with other parts that share its material nature and together with it constitute a system.

In short, the first way out of the difficulty would be to assert that semen is assuredly a homoeomery, but a homoeomery-2. Just as a horn, which is homoeomerous in respect of its material make-up, is, when its function is taken into account, inseparable from its configuration, so semen, which is homoeomerous in composition, would be inseparable from the *movement* on which its function depends. For, though it is a liquid, and consequently without a stable configuration, semen is nonetheless characterized by the internal motions that are proper to it, and which alone account for generation. It may further be noticed that Aristotle concludes the discussion in *PA* 2.9, 655b15–25 by adding:

> We have, on the other hand omitted semen (γονῆς) and milk from the study of homoeomerous fluids: for an examination of these belongs to the detailed treatment of generation. The one, in fact, is the principle that brings creatures into being, the other is what provides their nourishment at birth.[4]

This juxtaposition might be an indication that Aristotle is more or less consciously bringing semen closer to homoeomeries-2. It is not just a simple matter of emphasizing its functional aspect – in Aristotle's biology everything has a function. The real significance lies in distinguishing the functional autonomy of homoeomeries-2, which are subordinate only to the whole organism, from the narrow dependency of homoeomeries-1 in regard to a part that is anhomoeomerous. Thus, a horn, for example, is not a constituent of a head, but is designed to act as a protection for the organism as a whole. On the other hand, a cheek is a constituent of a face, in the sense that its function is an integral part of the more general

[4] Unless otherwise indicated, translations are my own.

function of the buccal apparatus. It is this that explains why a horn is a homoeomery-2, while a cheek is a homoeomery-1.

The second answer to the "weak *aporia*" is to reject the notion that semen is homoeomerous, on the ground that it is made up of two elements: a liquid envelope whose composition remains to be determined, but which there is every reason to regard as a homoeomery, and, enclosed within this envelope, bubbles of gas, or πνεῦμα.

Aristotle is still insufficiently explicit in *GA* 1.21, when he talks of a "power (δύναμις) and movement (κίνησις)" in the semen (729b5–6). The power is certainly described as active, efficient, and formative; it acts by means of heat (729b6–33). But we are given no specific information as to the material support underlying the generative principle beyond the fact that it is inherent in the semen. He returns to the subject of the composition of semen in *GA* 2.2. At the end of an account of its physical properties when exposed to heat or cold, Aristotle concludes that semen is "composed of breath and water," that is of "hot air" (736a1–2: θερμὸς ἀήρ) and water. Unfortunately, the word "composed" (συνιστάμενον) is ambiguous in Greek. It can just as easily refer to the result of a mixture – in which case we would have a classic homoeomery – as it can to a homogeneous juxtaposition, which would contravene the canonical definition of a homoeomery. It will not do to press the analogies Aristotle comes up with in this passage to demonstrate that semen consists of breath and water too hard. In fact he mentions five composite substances whose whiteness is due to the presence of air: foam, oil mixed with air, lead ore mixed with water and oil, snow, water mixed with oil. As far as one can see, in this context Aristotle feels not the slightest reluctance in speaking of a mixture, when in fact what he is talking about is a close juxtaposition. This point emerges clearly at least in the first illustration, his description of foam in 735b8–13:

> It is not only the fluid composed of water and earth that thickens, but also that composed of water and air (πνεύματος): for example even foam (ὁ ἀφρός) becomes thicker and turns white, and the smaller and more indistinguishable the bubbles become, the whiter and more compact the mass of it appears to be.

The question that immediately arises is therefore whether the hot air that explains the whiteness of semen is likewise present in the semen's water component in the form of bubbles, or whether the analogy does not cover this aspect of things. In short, we have no positive knowledge as to whether semen is a homoeomery, that is, the result of a mixture that satisfies the requirements of *GC* 1.10, or if it is a weak mixture, that is, bubbles in suspension in water that is very slightly earthy, in which case "for Lynceus nothing is mixed" (cf. *GC* 1.10, 328a15).

As his argument progresses, Aristotle's language makes it increasingly clear that what he has in mind is a weak mixture. After disposing of two misconceptions regarding, respectively, elephant semen and that of the Ethiopians, Aristotle returns once again to his theme (736a13–21):

> The reason for the whiteness of semen (τοῦ σπέρματος) is that the seminal fluid (ἡ γονή) is foam (ἀφρός) and that foam is something white, above all when it is formed from the tiniest of particles, so small that each bubble is individually imperceptible to the eye, as happens when water and oil are mixed together and stirred up, as was mentioned earlier. Even the ancients, it seems, did not fail to notice that semen is foam-like in nature. At all events, the name they gave to the goddess who presides over sexual union is derived from this property.

Aristotle's terminology is precise and univocal. He makes a distinction between semen and seminal fluid. It is the seminal fluid that is the foam. It is notable that Aristotle here avoids any of the comparative particles that he is so fond of elsewhere: the seminal fluid *is* foam. Now foam (or froth) has just been described as an emulsion of air-bubbles in a liquid. The bubble idea comes up once again and it is no accident that it appears in this context. Finally, the reference to the ancients would doubtless be somewhat extreme, if its sole purpose were to justify a remote comparison.

It might still, however, be supposed that Aristotle was, briefly, carried away by his own analogy. This skeptical line of defense is ruled out by one last passage whose importance there is no need to underline (*GA* 2.3, 736b33–737a1):

> There is, contained in the semen (ἐν τῷ σπέρματι) of all creatures, an agent that makes their semen (τὰ σπέρματα) fertile, namely what is called heat. Now this heat is not fire or any substance of that kind, but the breath stored up in the semen (ἐν τῷ σπέρματι) and in the frothy material (καὶ ἐν τῷ ἀφρώδει), and the natural property (φύσις) inherent in the breath, which is analogous to the element belonging to the stars.

In this passage, which of course presents all sorts of problems, Aristotle keeps to his "foam-like" model. The "breath" is contained in the frothy material, very likely in the sense that it is contained in the minute bubbles that constitute the foam. The breath itself, it may be argued, contains a "nature" (φύσις), and this might weaken the sense in which these things are literally "contained" one in another. I do not think so: while the bubbles are contained locally in the seminal fluid, the divine "nature" is inherent in them in the sense that it is a property, a virtue, an efficacity that resides in them. Whence the second answer: the "weak *aporia*" is resolved because the seminal fluid is not a homoeomery. Since the bubbles of breath can be of different sizes, and different concentrations in a given volume of liquid, their properties, at least in regard to configuration and size, will dictate the size and the number of viable embryos.

Should we choose between these two answers or find a means of accommodating both? My own preference is for the second course, for they are, as it were, only slightly contradictory. To be sure, the moment we take into account that seminal fluid is full of bubbles, it becomes clear that, strictly speaking, it is not merely not a homoeomery-1, but not even a homoeomery-2 either. How, after that, are we to take account of the passages in which Aristotle numbers semen among the homoeomeries, or even explicitly states that it *is* a homoeomery? I suggest that we should quite simply recognize a third sense of the word "homoeomery," closely parallel to the weak sense of the word "mixture," that is, mixture by juxtaposition of tiny particles. Given that Aristotle is content to call the result a "mixture" of oil and water when each molecule of the first is in contact with a molecule of the second, he would likewise allow this spurious type of mixture to be described as a homoeomery.

The question that now demands an answer is whether in the case of semen the homoeomery-3 that it constitutes comes closer to being a homoeomery-1 or a homoeomery-2. We cannot be certain. Insofar as semen is foam, it is a degraded form of homoeomery-1. But can semen be reduced to no more than just foam? In view of its prominent biological role, one may well doubt it. This vague impression is however still far from enough to justify a theory that semen should be classed as a quasi-homoeomery-2 (and not just as a quasi-homoeomery-1, which can from now on be taken for granted).

The Strong *Aporia*

This opening discussion brings us to the "strong *aporia*." It consists in wondering how the determination of the number and size of embryos works, if, as we have just established, semen is a homoeomery-3 – and that it should be the *one and only* homoeomery-3 that Aristotle has taken the trouble to characterize as a "homoeomery" proves that what is at stake is size.

We could restrict ourselves to a teleological answer: clearly the survival of individuals and thus of a species depends, as a litter of offspring confirms in any given case, on its satisfying certain conditions as to size and number. One might even try to include size among the considerations that belong to the form of a species. Thus, it would be proper to the human form for it to come into being within a spectrum that is not only qualitative but also quantitative. If a creature with the same qualitative structure were smaller than some minimum size or larger than some maximum, it would not be human. Two points to notice: this problem came to be seen as important from the time of Plotinus and through the Arabo-Latin Middle Ages to Galileo, but does not seem to have engaged Aristotle much. It is significantly present in the refutation of Anaxagoras' theories in *Phys.* 1.4, but the question begins to lose importance in the *GA*. Later, the problem is not sufficiently distinguished in Aristotle from that of the size of an individual. He only reestablishes a link with it again, so to speak, through the formal and final causes, but not through the cause that is the origin of movement or the material cause. In fact, what make an organism grow to maturity in size, are nutrition and the nutritive soul. Now when the semen imparts movement to the menses so as to produce an embryo, the process involved has nothing to do with nutrition. Even if, therefore, the two processes can be attributed to the same formal and final cause, the cause that is the origin of movement and the material cause are, on the other hand, two completely different things.

To explain the number and the size of the embryos in a litter, we cannot satisfy ourselves simply by pointing to the viability of an organism that will come from it; for the efficient cause and the material cause remain unaccounted for. Aristotle devotes large sections of *GA* 4.4 to this question. The first answer he provides is, on the face of it, the simplest. He relates the size of the full-grown creature to the volume of "seminal matter" from which it originates. Though Aristotle does not go into detail, it is clear that what he has in view is the portion of menstrual residue that receives its initial formation from the male principle. Here is the comparison Aristotle sets up:

> But, just as for animals when they are fully developed, there are upper and lower limits of size for each type, that determine how big or how small they can be, such that the excess or deficiency of size animals exhibit in comparison with one another falls within the range spanned by these limits, and a man is sometimes bigger, and sometimes smaller than another, and the same is true of any other animal – in the same way the semen-matter required for a viable embryo is not unrestricted (in which case any amount at all could produce a birth), but is subject to a maximum and a minimum limit. (*GA* 4.4, 771b33–772a4)

It is true that if an individual adult is to be of a certain size, it is easy enough to understand that the initial material of which it is made should have to possess a determinate size. But this first approach leaves the problem untouched; for the material cause is still not available, nor the efficient cause, to explain why this particular portion of matter is going to be singled out to receive the active principle. There can scarcely be much doubt that the "semen-matter" contributed by the female is undifferentiated, passive, and relatively inert. In that case, we cannot attribute to it a capacity somehow to acquire possession of the principle of movement, to confine it in a precisely delimited area, and to use it to transform itself into an embryo.

Aristotle makes the difficulty still clearer in the lines that follow:

> In consequence, for the reason stated, for all animals which emit more residue than is needed for the principle of one animal alone, it is not possible for only one embryo to come from it in its entirety, but just the number determined by the sizes that are appropriate, and no more will the semen of the male or the power contained in the semen make anything take on either more or less substance than is natural to it. (*GA* 4.4, 772a4–9)

But this last proposition can be understood either as a description of what is the case, or as a justification of what ought to be. An uncompromisingly teleological reading will opt for the second hypothesis. Semen would be naturally "programmed" to produce embryos from a given amount of matter in numbers either below or equal to the maximum of viable organisms possible. In other words, semen would, by itself, "recognize" the right quantity of menstrual matter to set in motion.

Truth to tell, the conclusion is inevitable. As we have seen, the menses are homoeomeries-1; semen is not a homoeomery-1, but a homoeomery-3 consisting of a homogeneous mixture of slightly earthy water and bubbles of hot air. This, then, would be the answer to our problem: the bubbles present in the semen are the only candidates that can conceivably provide the material and efficient causes needed to account for fertilization. This is what Aristotle says:

> [. . .] an embryo does not arise from just any quantity at all, but if there is <too> little of it, or much too much, there will be no embryo. In fact the capacity of the passive element and that of the heat that acts upon it has definite limits. (*GA* 4.4, 772a27–30)

We are not to believe that it is enough for an embryo to form that the semen should bring in a certain quantity of heat. Earlier in the treatise, Aristotle laid considerable emphasis on the movement impressed upon the menses. The "heat that acts upon it" must assuredly possess a "definite limit," but not simply *qua* thermal property; the dynamic aspect of fertilization is essential. Unfortunately, Aristotle nowhere gives an account of it. To make good for this lack, it seems worth investigating what our sources have to say on the subject of bubbles.

Bubble and Fermentation in *Timaeus* and *GA*

Aristotle is not the first to bring the notion of "bubble" into natural philosophy. Anaxagoras may possibly have mentioned the subject. It is in particular to be found appropriately located in the *Timaeus*, in contexts that clearly form part of the background to Aristotle's treatment in the *GA*.

Thus, Plato at 66 A 2-B 7 writes:

> Again, as to those substances that are rendered volatile by putrefaction and make their way into narrow veins, being proportionate to the earthy particles to be found there and to all the air that is contained in the veins, with the result that, by being moved round each other, they produce an emulsion, and once the emulsion has been formed, they spurt up all round and, penetrating throughout it, they repeatedly produce hollows that enclose what comes into them – substances, then, that, (given that a hollow of moisture, sometimes earthy (γεώδους), sometimes clear, encloses the air), become moist receptacles of air in the form of round (περιφερῆ), hollow droplets, which, in the case of the transparent envelopes made from clear moisture, are called "bubbles" (πομφόλυγας), and where the moisture is earthy and whipped up into a coherent mass, are referred to as "efferves-cence" and "fermentation" – well, the agent responsible for these effects is given the name of "acid."

This translation, unlike those usually offered, preserves the length and tortuousness of Plato's periodic style, because this matters for the author's philosophic plan. It is part and parcel of the program of the *Timaeus*, which is intended as an in-depth exploration of the complexities of the world of sense, to adopt a syntax that reflects the obscurity of the subject. That Plato should devote a sentence of such length to the description of a bubble is particularly delightful. The theory he develops is just what we find in Aristotle. A bubble is made up of air imprisoned within a hollow liquid surface. For it to be truly a bubble the liquid must be clear. Otherwise what we have is effervescence or fermentation. It is of some interest to note that Plato does not say precisely what the substances are that are "rendered volatile by putrefaction." There can be little doubt that they present themselves in the form of a liquid of some sort, since these are what are soon going to enclose the air present in the veins. If this is so, we see that the bubble is "round" (περιφερῆ) but not circular in a geometric sense. In reality, from the fact that it is, like all bodies, composed of elementary solids (the liquid result of putrefaction and perhaps the particles of earth that it finds on the spot), its surfaces are rectilinear. This point will require further discussion.

Plato returns to the bubble a little further on, in his description of white phlegm. He brings it in first at 82 C 7-E 1:

> Again, the liquid matter that results, along with air, from the decomposition of young tender flesh, when it is inflated by wind (ἀνεμωθέντος) and enveloped by moisture, and when, as a result of this, bubbles are formed, *individually invisible because of their smallness, but collectively producing a visible mass with a color that is white in appearance through the creation of foam* (καθ' ἑκάστην μὲν ἀοράτων διὰ σμικρότητα, συναπασῶν δὲ τὸν ὄγκον παρεχομένων ὁρατόν, χρῶμα ἐχουσῶν διὰ τὴν τοῦ ἀφροῦ γένεσιν ἰδεῖν λευκόν), all this combination of putrefied, tender flesh and air we call "white phlegm." The secretions that come from newly formed phlegm are sweat and tears which, along with all the other bodies of this sort, are purged daily as liquids.

The description of the whiteness of the emulsified body is almost word for word what turns up in Aristotle. Plato's bubbles (πομφολύγων), described as "individually invisible because of their smallness, but collectively producing a visible mass with a color that is white in appearance through the creation of foam," are the direct precursors of Aristotle's in the crucial passage at *GA* 2.2, 735b8–13:

> It is not only liquid composed of water and earthy material that becomes thickened, but also that composed of water and breath (πνεύματος); for instance, foam also (καὶ ὁ ἀφρός) thickens and turns white, and the smaller and more indistinguishable (ἐλάττους καὶ ἀδηλότεραι) the bubbles are, the whiter and more compact is the appearance of the mass (ὁ ὄγκος) of it.

Thus Aristotle takes over from Plato not only the general explanation of the whiteness displayed by emulsified liquids, but the particular example of foam and the terminology used in the *Timaeus*. This indeed makes the difference in the way Aristotle uses it all the more striking. Plato's description of white phlegm is taken over *almost* as it stands to describe semen. The fundamental difference is of course that Aristotle substitutes "breath" (πνεύματος) for Plato's "air" and "wind." One wonders, frankly, if Aristotle is not employing a verbal shift; for he brings in "breath," not in order to differentiate semen from other white emulsions (which in any case would contain bubbles of *air* not *breath*) but to describe foam. At this stage in the argument, "breath" and "air" therefore appear interchangeable.[5]

Plato mentions bubbles in a third passage, in which he returns to the subject of white phlegm, *Timaeus* 85 A 1-B 2:

> White phlegm is dangerous because of the bubbles, if it is trapped inside the body; if however it escapes as exhalations outside the body, it is milder indeed in its effects but it disfigures the body in various ways by giving rise to white eruptions and other related ailments. If it is mixed with black bile, and dispersed toward the revolutions in the head that are most divine so as to upset them, it is gentler when it invades during sleep, but much harder to shake off if it attacks people who are awake. The illness is very properly called the "sacred disease" because it affects a part that is sacred in nature.

This passage is interesting on a number of different counts. At the beginning of the well-known *Problem* 30, the author mentions the "sacred disease" (30.1, 953a16), but he associates it with black bile without particularizing further. It is then, apparently, an innovation of Plato's to make white phlegm the vector of the black bile toward the "revolutions in the head that are most divine." In so doing, Plato is giving credence to the theory that emulsified liquids, here assimilated by him to "exhalations" (ἀναπνοάς), have dynamic potentialities. It was but a short step from there to attribute an internal capacity for movement to semen.

If we sum up the main features of the bubble theory that Aristotle was able to find in the *Timaeus*, the following five points emerge:

- sphericity;
- different possible envelopes, more or less clear;
- different possible sizes;
- macroscopic whiteness of the liquid resulting from microscopic *gaseous* pockets;
- internal motor functions.

In the *GA*, bubbles make their appearance at two strategic points, as the vehicles of the heat present in sexual generation and spontaneous generation. The envelope that constitutes a bubble is treated in two different ways, and we may wonder if this difference is not implicit in Plato's distinction in 66 A–B between "bubble" and "fermentation." The first

[5] This indecisiveness is a clear indication of the tension between a basic definition of "breath" as "hot air" (736a1–2), and an elevated description of it as a body "other [. . .] and more divine than what are called the elements" (736b30–31).

passage to be taken into consideration is the one where Aristotle gives his account of the material substance of souls:

> But just as souls differ from one another in their relative value, or lack of it (ὡς δὲ διαφέρουσι τιμιότητι αἱ ψυχαὶ καὶ ἀτιμίᾳ ἀλλήλων), so such substances also differ in nature. (*GA* 2.3, 736b31–33)

The body "analogous to the element belonging to the stars" may thus be superior or inferior in quality, which explains the difference between the souls of the different sexed species. Aristotle returns to this device further on:

> The body of semen which serves as a vehicle to transmit the substance of the soul-principle, of which one aspect is separate from the body in the case of animals that have something divine in their make-up (what we call "intellect" is of this kind), while its other aspect is inseparable – this element of the semen dissolves and evaporates because it has a moist, watery nature. (*GA* 2.3, 737a7–12)

There is no doubt that Aristotle is here concerned with the question of the persistence or otherwise of the liquid medium containing the bubbles. He does not in any way address the question of what becomes of the "more divine" body confined in the bubbles, but considers only whether the liquid is incorporated in the menses in the making of the embryo. He answers in the negative. The liquid enveloping the bubbles "dissolves and evaporates." If Aristotle does not ask himself the same question in regard to the "more divine" body, it is simply that the question of its persistence does not arise. Since the material of which it is made is quite insubstantial, it follows automatically that it does not form part of the material composition of the embryo. At the very most, it merges with its generative power. Either, then, it will subsist as an attenuated gas mixed in with the flesh of the embryo, or it will be entirely dissolved. In other words, for embryos produced by sexual fertilization, the matter of the bubble is a liquid pure enough to disappear without leaving any trace on the embryo. It is, then, a "bubble" in exactly the restricted sense of the *Timaeus*, that is to say, a bubble as opposed to a "fermentation."

Now let us look at the account of spontaneous generation in *GA* 3.11, 762a18–32:

> Animals and plants are formed in the earth and in moist conditions, because in the earth there is water, and in the water there is breath, and in all breath there is soul-heat, so that, in a way, all things are full of soul; that is why they take shape rapidly, once it has been enclosed; it becomes enclosed, that is, it comes into existence, when moist corporeal substances heat up, as a frothy bubble. The differences that determine whether the

> kind of creature formed is of greater or lesser worth (αἱ μὲν οὖν διαφοραὶ τοῦ τιμιώτερον εἶναι τὸ γένος καὶ ἀτιμότερον τὸ συνιστάμενον) reside in the enclosure of the soul-principle, which in turn depends on the location in which it takes place and the corporeal matter enclosed. Now there is an abundance of earthy matter present in the sea. That is how the mollusks that are formed in such conditions come to acquire their particular nature: earthy matter hardens all round them, solidified in the same way as bones and horns (for they are not liable to melt from the effects of fire), and on the inside the body that possesses life is enclosed.

The subtle parallel between the two accounts is instructive. In both cases, Aristotle has recourse to the bubble notion to explain generation. But whereas, for sexual generation, the clear liquid that forms the bubble envelope plays no material part in constituting the embryo; for spontaneous generation, it becomes a material constituent of the creature formed. With Aristotle's insistence, in this case, on the presence of an *earthy* element (γεῶδες) widely diffused throughout the liquid, we are in exactly the same situation that Plato designates as fermentation, as opposed to bubble (*cf.* 66 B 1 and 4: γεώδους).

The parallels between the two accounts do not stop there. We should also notice how the content of *GA* 2.3, 736b31–32: *just as souls differ from one another in their relative value, or lack of it* (ὡς δὲ διαφέρουσι τιμιότητι αἱ ψυχαὶ καὶ ἀτιμίᾳ ἀλλήλων), is echoed in almost identical terms in *GA* 3.11, 762a24–25: *the differences that determine whether the kind of creature formed is of greater or lesser worth* (αἱ μὲν οὖν διαφοραὶ τοῦ τιμιώτερον εἶναι τὸ γένος καὶ ἀτιμότερον τὸ συνιστάμενον). By using the same vocabulary Aristotle draws attention to the same point, namely the differences in worth, that is, in value between the various souls. In the case of sexual generation, the difference is uniquely explained by the nature of the "more divine" matter, since the liquid disappears once the heat has been conveyed to the matrix. In the case of spontaneous generation, on the other hand, it is the envelope itself that does duty as the matrix. The quality of the offspring will therefore be determined by the quality of the "enclosed" breath – as for sexual generation – *and* by that of the envelope that performs the matrix function. We might draw the conclusion that Aristotle has neglected to mention the quality of the menses, but that he would not have denied that this also plays a part in the value of the offspring. In truth, the point is so obvious that it did not need to be stressed.

Here we can investigate a first genealogical conclusion: Aristotle sticks closely to the bubble theory that he finds worked out in two passages in the *Timaeus*. Does this mean that Aristotle is still a Platonist in his theory of generation? No, for two reasons: one of them is obvious but quite superficial, the other diffuse but more profound. Here is the first: even if he adopts Plato's physicochemical model, virtually as it stands, Aristotle diverts the bubble's function in order to apply it to the generation of animals. The theory of generation he develops is then, evidently, original. At a deeper level, we have already alluded to the fact that the Platonic bubbles, even if they are "round" (περιφερῆ), could nonetheless not be geometrically spherical, since they are composed of water, that is, of a multitude of juxtaposed icosahedra. At a microscopic level, Platonic bubbles are, then, made of elements characterized by solid, rectilinear angles. There is nothing like that in Aristotle. Although, in the *GA*, he does not liken the shape of a bubble to that of a geometrical sphere, it is clear that this is the model he has in mind. The author of section 16 of the *Problems* (entitled *On Inanimate Objects*) is, on this point, evidently faithful to the spirit of Aristotle's text. At *Problems* 16.2, 913a29–33, he writes:

> Why are bubbles hemispheres (διὰ τί αἱ πομφόλυγες ἡμισφαίρια)? Is it because they are, as from a center, carried upwards to the air, equally in all directions? That necessarily produces a hemisphere. But the lower hemisphere is cut off by the surface of the watery body on which the center lies.

Segment of a Circle and Hylomorphism

The sphere is, of course, a fully fledged object in Aristotle's physics and cosmology, which is not the case in Plato's. But it is also one of the canonical examples Aristotle uses to illustrate hylomorphism. In particular, it figures strongly in *Meta.* 7 at strategic points. The textual connection is guaranteed by a passage in our treatise, *GA* 1.21, 729b14–18:

> So that, if we take the male and the female in their strictest sense, that is, the one as having productive and motive capacity, and the other as capable of being acted upon and moved, then what is produced from them will not be one, except in the sense that a bed is the product of a carpenter and a sphere is made from wax and the form (ἧ ὡς ἐκ τοῦ κηροῦ καὶ τοῦ εἴδους ἡ σφαῖρα).

This is the subject of the excursus in *Meta.* 7.7–9 concerning natural, artificial and spontaneous genesis. In *Meta.* 7.8, in particular, Aristotle returns at some length to the example of the bronze sphere to explain that, in this composite object, the form is not something that could be produced any more than the material is (1033b5–10). Let us suppose that a craftsman, on a given occasion, produces the bronze sphere. It will then exist because it preexisted in his mind. That, again, is only possible because there exists an anterior form which he derives his inspiration. To put it briefly, form is never created, because it has always existed. This does not mean that it has to have the kind of eternity possessed by the Platonic Ideas, but only the sempiternity of Aristotelian forms. This is the context that seems to me to explain the controlled shift in *GA* 1.21 between the bed example and that of the sphere. It looks very much as if Aristotle had mentioned the formal cause of the wax sphere instead of the active and motive cause. This shift confirms that, for Aristotle, the formal cause is present "in" what acts so as to be able to end up in what is acted upon. In *Meta.* 7.10, Aristotle's account becomes more complex. He aims to show that the definition covers a compound of matter and form, but that it only takes account of the parts of the form. This subtle nuance leads Aristotle to a closely argued discussion of the notion of *part*, in its relationship to the parts of the compound that are themselves matter and form. Aristotle makes a distinction in the case of the bronze circle, between two types of matter, the bronze and the "segments" of the circle (*Meta.* 7.10, 1035a9–14).

> This is the reason why the definition of the circle does not include that of the segments, while the definition of the syllable includes that of the letters; for the letters are parts of the definition of the form, and not <the> matter <of the syllable>, while the segments of the circle are parts in the sense of material parts on which the form supervenes, although the segments are nearer to the form than the bronze is, when roundness is produced in bronze.

So far, nobody appears to have offered a satisfactory explanation of these "segments of a circle" (τμήματα) that recur over and over again throughout 7.10. Basically, the first question to be asked is what their function is. Why, at the material level, make a distinction between bronze and the segments of a circle?

I would like to suggest that Aristotle's motive is to defend his hylomorphism and use it to construct a telling argument against Plato's realism. His defense of hylomorphism consists in distinguishing several material levels, so that the objects of mathematics can be shown to exist on a plane that is not distinct from the sensible.[6] For if the only line of demarcation between matter and form separated the bronze (sensible) from the circle (mathematical), the Platonists could consider themselves justified in claiming that the circle as such is not sensible. Alike in respect

[6] See Ross' note in Ross 1958, vol. 2: 197 (*ad* 1035a20).

of its geometrical matter – its extension properly speaking – and in respect of its geometrical form – its circularity – the circle would belong in another level of the real, rightly based on the uniqueness of its geometrical matter, *viz.* extension. To meet this objection, Aristotle wants to untie the bond that may exist between intelligible matter, that is, geometrical extension and geometrical form, in this instance circularity. Aristotle, then, makes it clear that geometrical matter should be put down as just simply matter, that is, assigned to the basic world of sense, and not to that of the form that resides in it. Intelligible matter does not, as such, constitute another plane of reality; it is merely an aspect of matter full stop, one way of looking at it. As Aristotle puts it a little further on:

> Matter is in some cases sensible and in others intelligible; sensible matter includes for example bronze, wood and all matter that is subject to change, intelligible matter is that which exists in sensible things not *qua* sensible, in the way that is true of the objects of mathematics. (*Meta.* 7.10, 1036a9–12)

It will be objected that in general situations of this type, where the intuition on which one philosophy is founded stands in stark opposition to that of another – in this instance, where the Aristotelian theory of abstraction comes along to contradict the Platonic theory of intuition – it is not so much a question of true and false as a matter of philosophic taste. Aristotle's criticism is more nuanced. It seems, in effect, to find fault not so much with the general line taken by the Platonists' ontology as with their very marked incapacity to account for change in the sensible world and, quite particularly, for generation. In the world of Platonic geometry, in which geometrical configurations exist by themselves and not in what is perceived by sense, it is impossible to distinguish their size. The size of a circle belonging to a purely geometrical space will be, at most, a matter of convention, in that a certain diameter will be adopted as a unit of comparison, and circles will be arranged according to the relationship of their diameter to the diameter-unit. Geometrical space eschews absolute size and admits only relative sizes. Consequently, it is impossible to understand how a geometrical form could "produce" a sensible object of similar form, which would never be simply a form, including a "materialized" form, but would always be, for the very reason that its existence is sensible, a form that can be quantified in absolute (and not simply relative) terms.

This handicap to participation would be acceptable if we were only concerned with rectilinear polygons. So let us take the elementary triangles in the *Timaeus*. Their characteristic is that they can be broken down into parts that are identical triangles. This equivalence in shape between one form and its constituents, ad infinitum, is principally an opportunity for Plato to neutralize the question of their size. In the end, it all comes down to the same triangle, the same triplet of angles. Whatever size of half-equilateral triangle or half-square I choose it will always form the *same* angle with an external straight line. Or again, the three angles of a "little" triangle will be exactly superposable, in pairs, on the three angles of the "big" triangle. This is not the case with circles. It is in the light of this that the repeated mention of the segments of a circle must be understood. In particular, Aristotle says at *Meta.* 7.10, 1035a31–b3:

> And for this reason the clay statue decomposes into clay and the sphere into bronze, Callias into flesh and bone, and again the circle into segments – for there is a kind of circle which is associated with matter. In fact, the word "circle" is used ambiguously, for circle in an absolute sense and for individual circles, because individual circles have no name of their own.

The last three passages cited prove that Aristotle sees a particular connection between the "intelligible matter" (ὕλη νοητή) of the circle and its "segments" (τμήματα). This connection, despite its great intrinsic interest, does not appear to have been identified by commentators either ancient or recent. The hypothesis I should like to suggest here is that the segments, i.e., the arcs of the circles, are the intelligible matter in that such matter is, precisely, not dissociable from sensible matter, but is only an aspect of it. It is what is left once the sensible matter has had its properly corporeal nature abstracted from it. Take the bronze away from the bronze circle and we still have, in this particular precise part of the world we inhabit, a circle that is quantifiable. In other words, the geometrical object exists nowhere else but *in* the sensible. Now, and this is the second act in Aristotle's critique aimed at the Platonists, this ontological demotion is offset by the possibility of assigning a definite size to the geometrical object. The geometrical circle in the sensible object, viz. the circle obtained by abstracting the bronze from the bronze circle, is a configuration furnished with an absolute measure.

This interpretation appears to be confirmed by the proof Aristotle provides, in *APr* 1.24, of the equality of the base angles of an isosceles triangle, in the course of a demonstration of the generality necessary to the syllogism. He writes:

> Let lines A and B be drawn <from a circumference> to a center. If you assume that angle A+Γ is equal to angle B+Δ, without postulating in general that angles inscribed in semicircles are equal; if, in turn, you assume that angle Γ is equal to angle Δ without further postulating that all angles

inscribed in the same segment of a circle (τοῦ τμήματος) are equal; if finally you postulate that when equal angles are subtracted from whole angles that are themselves equal, the angles remaining [E and Z] are equal, you will be guilty of a *petitio principii,* unless you postulate <in general> that when equal angles are subtracted from equal angles, the angles remaining are equal. (*APr* 1.24, 41b15–22)

With some changes to what the letters designate to make the proof more explicit, the proof can be illustrated by the following (Figure 1):

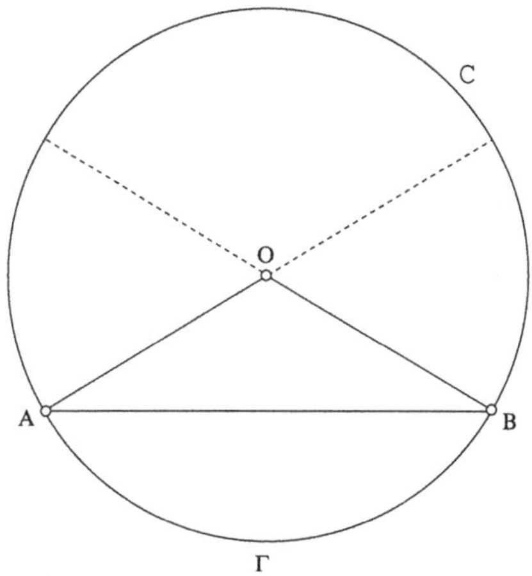

Figure 1.

Though this passage has not traditionally been invoked in the interpretation of *Meta.* 7.10, it appears to throw considerable light on its meaning. According to what seems by far the most likely interpretation put forward by the ancient commentators, Aristotle is offering a proof involving mixed angles, made from a straight line and the arc of a circle. In this context, he uses the term "segment" (τμῆμα) for the portion of the circle cut off by the chord, and "angle of the segment" (τὴν τοῦ τμήματος, sc. γωνίαν) for the angle formed by the chord and the arc of its circle. As Ross notes, following Heiberg: "The use of mixed angles had probably played a larger part in the pre-Euclidean geometry with which A[ristotle] was familiar, though comparatively scanty traces of it remain in Euclid."[7] The present passage, therefore, furnishes the proof that Aristotle associated the notion of the cornicular angle with that of the "segment" of a circle.

[7] Ross 1949: 375.

This pre-Euclidean proof of the equality of the base angles of an isosceles triangle is, however, marked by a difficulty which might well have confirmed Aristotle in his philosophic choice of hylomorphism. Take Figure 1 and describe on it a circle C′ with center O′ and radius O′A, with O′ on segment OA. Segment AB cuts this circle at B′ (Figure 2).

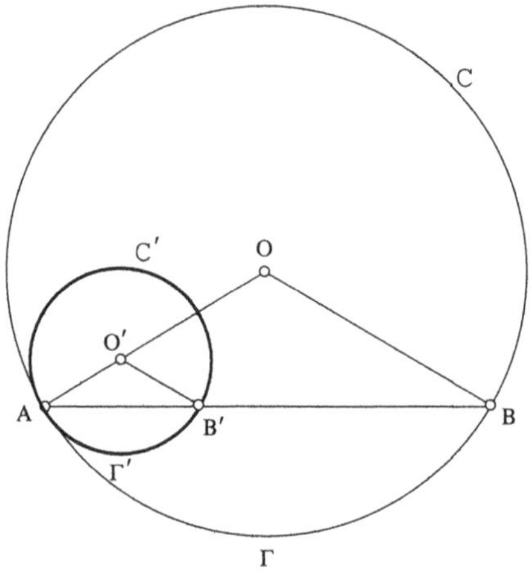

Figure 2.

By construction, angle O′AB′ is identical to angle OAB. As, by construction, triangle (O′AB′) is isosceles, angle O′B′A is equal to angle O′AB and thus to angle OBA. In virtue of the similarity of the figures, the mixed complementary angles, under the chord, of the two base angles of an isosceles triangle are equal. Consequently, the mixed angle B′AΓ′ is equal to the mixed angle BAΓ. Now, in virtue of the definition of the contact between two internally tangent circles, our two circles only touch each other at the single point A. It follows that the two angles "of the segments" B′AΓ′ and BAΓ are equal, but are produced by two arcs of a circle that do not cover one another. In other words, the cornicular angle is not properly speaking constituted by the arc of a circle: it is an infinitesimal potentiality *at the birth*, as it were, of the arc of a circle.

As he shows in the *APo* passage, Aristotle accepts this concept embraced by the mathematicians of his time, but he confines it to the purely operational sphere of mathematics. The arcs of the circles containing these paradoxical angles are a quite different matter: these arcs have the peculiarity of not covering each other (Figure 3, left). On the other hand, in the case of the rectilinear figures, the only ones Plato admits, the equality of the angle (Figure 3, right) entails the superposition of the segments. The question of size is therefore removed.

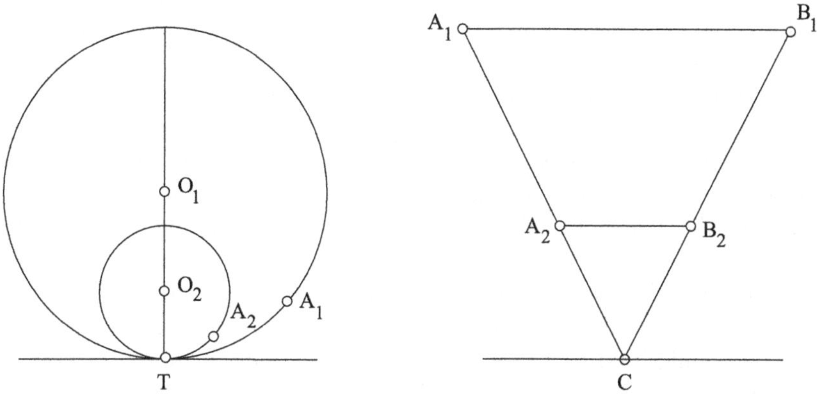

Figure 3.

This nonsuperposition of the arcs of a circle gave Aristotle a decisive argument against Plato, and it is this, in my submission, that explains why he keeps mentioning "segments" in the way he does in *Meta.* 7.10. The ideal status Plato accords to geometrical figures cannot accommodate the paradox that we have just brought out, if he accepts that a circle is a geometrical figure just like any other. The problem of the angular equality coupled with the impossibility of superposition characteristic of the angle of the segments can only be explained, or rather neutralized if, in one way or another, the Ideas are "incarnated" in the sensible – whether in Aristotle's way or that of Eudoxus – or the circle is denied the status of a geometrical figure – which, as I have suggested elsewhere,[8] might represent the actual position Plato assumed, thus isolating himself from the rest of his own school.

Be that last point as it may, Aristotle appears to act as if his adversary recognized the existence of the circle as a geometrical figure. Platonism, which explains forms, or relative magnitudes, but not absolute magnitudes, cannot account for the complete absence of superposition at point A of circles C and C′ (Figure 2). Hylomorphism, on the other hand, by recognizing no circles but those that effectively exist in the sensible world, is content to take note of the paradox without having to bear the burden of a proof of their existence.

[8] See Rashed 2013: 215–31 (especially 224–5).

Answer to the Strong *Aporia*

In the "dynamics equations" at the end of *Phys.* 7.5 (250a10–25), Aristotle admits the existence of thresholds breaking the simple proportionality of motive force, weight, distance, and time. If, then, we admit a direct relationship between the quantity of vital heat and the size of our bubbles, we can understand how the emulsive quality of semen – i.e., the fact that it contains bubbles of adequate size – could be a *condicio sine qua non* of fertilization. But add a further hypothesis: suppose that, as would be natural, the semen bubbles were extremely numerous and of different sizes. They will then have different arcs, different τμήματα. If, as Aristotle is fond of saying, semen behaves like a craftsman, we can compare this multitude of bubbles to all the tools in his toolbox. The difference between the arcs could hold part of the genetic information which, up until now, we did not exactly know where to locate.[9]

The analogy between the circle and the sphere, with due allowance for the paradoxes inherent in the cornicular angles, had no doubt been accepted since the time of Democritus. The constant oscillation between "circle" and "sphere" in *Meta.* 7.10 proves that Aristotle is conscious of the difficulty as well as the legitimacy of the extension from two to three dimensions. The attribution of the responsibility for the process of generation, sexual and spontaneous alike, to two types of bubbles – something Plato did not do, in spite of his interest in these two types of bubbles in the *Timaeus* – and the adoption of a hylomorphist ontology go hand in hand. Because the bubbles are mathematical spheres in the Aristotelian sense, and because the sensible spheres have sizes that are peculiar to them (τμήματα), the active principle which, coming from the male meets the undifferentiated seed of the female, is provided with a sufficient range of individual configurations, in sufficiently determinate sizes, to explain how an equally individualized and determinate amount of menstrual matter acquires a form, and offspring, determinate in number and size, are born from this process. In conclusion, semen has all the characteristics of an organ. What distinguishes it from the heart and the viscera is not so much that its nature is nonhomoeomerous as that its "weak mixture" is not amorphous like the "strong mixture" of the menses, but dynamic, active,

[9] Averroes seems to have made a similar point in his *Middle Commentary on GA* 2. See, in particular, f. 86 K14–L4: *nam hac portione aerea tantum potest formare figuram membrorum extrinsecus et facere concavitates eorum intrinsecus, sicut facit vitrearius figuram vitri, inflando ipsum.* On this text, see Cerami 2015: 509.

and linked to a particular configuration, like those other parts to which Aristotle expressly assigns a mixed status between homoeomerous and anhomoeomerous.

It is unfortunate that no text has been preserved in which Aristotle spells out the importance of the bubble in his biology. The subject did not, however, escape the attention of Theophrastus, who mentions the bubble in a key passage of his *Metaphysics*. After criticizing those who maintain that the basic principles are indeterminate but that the cosmos derived from them is a model of orderliness – very likely Democritus is the target[10] – Theophrastus raises the following difficulty against those who hold that the principles are determinate:

> On the other hand, it is difficult to invest each thing with a reason linking it up with its final cause in all cases, both in animals and in plants and in the bubble itself (καὶ ἐν αὐτῇ πομφόλυγι). (*Meta.* 7a19–22)

This point is highly intriguing. In fact it is aimed principally at the elementary physical theory advanced in the *Timaeus*, which would find it all the more difficult to explain a bubble because the basic principles of perceptible matter it postulates, triangles, are all rectilinear. But it might also suggest that Aristotle's own theory, even if less improbable than Plato's, is nonetheless still difficult. That Theophrastus had this biological context too in mind seems to me confirmed by what follows:

> Unless it happens that all sorts of shapes and varieties of things in the air and on the earth arise because of the order and change of other things. Some indeed take as the greatest example of this process, the phenomena linked to the seasons of the year within which the generation of animals, plants and fruits takes place, the sun acting, so to speak, as their begetter. (7a22–b5)

It does not seem to me to be due simply to chance that Theophrastus cites "the bubble itself" in a context so strongly marked by biological references. What we have is nothing else but an implicit exegesis of Aristotle's embryology, which held that if it is indeed a "bubble" that is responsible for reproduction, it is a bubble "analogous" to the "element of the stars," and so, probably, capable of being influenced by their periods.

[10] Theophrastus, *Meta.* 7a9–19.

CHAPTER 7

Function and Instrument
Toward a New Criterion of the Scale of Being in Aristotle's Generation of Animals

Cristina Cerami

Introduction

In several places in the biological corpus, Aristotle declares that, among living beings, some animals are more perfect (τελειότερα) and more valuable (τιμιώτερα) than others. At the beginning of *GA* 2, he identifies the most valuable animals with those that have a hotter and moister nature. These two claims, however, are problematic for at least two reasons.

For one thing, the very idea that different species of animals can be classified according to a normative criterion is itself far from straightforward. In fact, one of the main tenets of Aristotle's ontobiology is that each species, in so far as it is the expression of a certain essence, is an independent class of beings that neither exists for the benefit of another nor can be considered "that for the sake of which" other species exist. In this sense, it is not easy to understand how Aristotle can even admit the possibility of establishing a classification according to which some species are *as such* better than others.

For another thing, one could infer that, since the classification is established according to the amount of heat in the constitution of animal species, heat, or more precisely vital heat, has a real explanatory function in the generation of animals and in their activities. Such a contention might call into question the fundamental grounds of teleological explanation, which entails that the generation and the development of living organisms cannot be wholly explained by the simple natures and potentials of their material elements, without reference to the form or nature of the living organism as a whole.

This essay is a revised version of a presentation given at a conference on Aristotle's *GA* organized by David Lefebvre and Allan Gotthelf. The texts I discuss in this essay are also discussed in more detail in Cerami 2015. I am grateful to David Lefebvre and Andrea Falcon for their remarks and suggestions on an earlier draft.

This essay tackles this major issue in Aristotle's theory of animal generation. By reconsidering some crucial passages from Aristotle's writings on natural philosophy, I will suggest that these two difficulties have a common solution that lies in the way one defines living beings *qua* living beings and what is good for them. In building on some of Allan Gotthelf's most insightful suggestions, I will argue that neither the quantity nor the quality of heat in living beings provides the ultimate criterion for the hierarchical order envisioned by Aristotle.[1] It is rather the function (ἔργον) that the vital heat makes possible – to be identified as the immanent goal of a living being – that determines this order. In this context, it will be of major importance to examine how far this same criterion can be applied not only to living beings but also to inanimate entities and, more generally, to nature as a whole. I hope to show that even if this ἔργον is something common to all natural beings only by analogy, it forms the basis of a unique and continuous scale of being.

Scale of Perfection and Quantity of Heat

Throughout the *GA*, as well as in other passages of the biological corpus, Aristotle declares that living beings are not all the same, since some of them are more valuable and more perfect than others. This is true not only of the different species, compared one with another, but also of the different individuals within each species. A species can be more valuable than another, an individual more valuable than another and, notably, male individuals are more valuable than female ones. Even if Aristotle does not clarify the criterion of the hierarchy, he repeatedly declares that nature orders beings and their generation in regular gradation, so as to establish a continuous scale of being.[2]

On this picture, *GA* 2.1 is a particularly crucial text, since Aristotle considers three different classifications in which some members are more valuable than others but includes these classifications within the scope of one single inquiry into the causes of the differentiation between the female and male individuals in the species in which this difference exists.[3]

At the beginning of the chapter, Aristotle declares that a complete explanation of the coming into being and the existence of male and female

[1] Gotthelf 1987 (reprinted in Gotthelf 2012a). A similar argument is found in Balme 1987b.

[2] *GA* 2.1, 733a32; cf. *HA* 8.1, 588b4–589a9, *PA* 2.10, 656a7 ff., *PA* 4.5, 681a12–28.

[3] For a discussion of this stretch of text in the context of the overall argument of *GA* 2, I refer the reader to Chapter 3.

must be given both on account of what is necessary, i.e., the agent and material causes, and of "what is better (βέλτιον) i.e. the final cause."[4] He makes clear that the explanation on account of what is better is derived from "a principle from above" (ἄνωθεν ἔχει τὴν ἀρχήν). He argues that it is better that the male principle is separated from the female one, since as far as possible what is better has to be separated from what is worse, and concludes that the male principle is better because it is the principle of movement.[5]

Afterward, in pursuing this line of reasoning, Aristotle takes into consideration the case of animals which emit semen and suggests that those animals are more valuable (τιμιώτερα) and more self-sufficient (αὐταρκέστερα) owing to the amount of vital heat (θερμότης ψυχική) in their constitution. Actually, since heat is a motive force (τὸ δὲ θερμὸν κινητικόν) and motion – one must infer – is what makes animals self-sufficient, animals that are hotter (these animals are also those that emit sperm) are more valuable, more self-sufficient, and for the same reason bigger than others. Living beings of this kind, concludes Aristotle, are blooded animals.[6]

From the above considerations, we can infer that species or individuals that are endowed with a stronger motive principle must be considered more valuable than others, since they contain a greater amount of vital heat, which is a motive force. This hypothesis is confirmed in several passages of the biological corpus where Aristotle states that male individuals are more valuable than female individuals, in that their constitution is hotter than the female one.[7]

In the following lines of *GA* 2.1, however, Aristotle seems partially to adjust this conclusion, by making clear that neither motion as such nor the organs of locomotion provide a means of differentiating living beings *qua* living beings, or the real criterion of their normative classification. After distinguishing the different classes of living beings according to their mode of reproduction, he concludes that, among them, one must consider live-bearing animals to be the animals more perfect in their nature (τὰ τελειότερα τὴν φύσιν) *and* (καί) participating in a purer principle (μετέχοντα καθαρωτέρας ἀρχῆς).[8]

We understand thus that the beings that Aristotle includes in the classifications are the living beings as far as they are capable of reproducing themselves. For in the three cases considered (i.e., the male as compared

[4] *GA* 2.1, 731b20–23. [5] *GA* 2.1, 732a1–11. [6] *GA* 2.1, 732a11–23.
[7] On this point, see Lefebvre 2014. [8] *GA* 2.1, 732b15–29.

with the female, blooded animals as compared with other animals, and live-bearing animals as compared with the rest of living beings), the members of the different classes are analyzed according to their contribution in reproduction.

In the three cases, moreover, the amount of hot matter and its quality seem to be the factors determining the excellence of one class over the other. For, in all three cases, heat is defined as the principle responsible for the better constitution of the more valuable class of animals. The same state of affairs is confirmed immediately after, when Aristotle affirms that the more valuable (τελειότερα) animals are those that have a hotter nature (θερμότερα τὴν φύσιν), have more moisture (ὑγρότερα), and are not earthy (μὴ γεώδη) in their composition.[9]

Aristotle's explanation of spontaneous generation, that is, his explanation of the generation of those animals which come to be neither from semen nor from a genitor of the same species or genus is further evidence of the major role that heat or, more precisely vital heat, plays in fixing the degree of excellence of a species. In offering an explanation that extends some of the results reached in the study of sexual reproduction, Aristotle points out that heat has a fundamental role in spontaneous generation for at least two reasons: not only because the residue out of which the new being comes to be is the product of a "concoction,"[10] but also because the new form of life appears when the *pneuma*, which is permeated by vital heat, is enclosed in a parcel of water or earth, so that, as the corporeal liquids become heated, there arises a "frothy bubble."[11] At this point, Aristotle declares that even in the case of animals of this kind we can establish a hierarchy of excellence and classify them according to a normative criterion. He goes on to explain that, among beings spontaneously generated, the hard-shelled animals (ὀστρακόδερμα) have a hard shell that protects the life-containing body enclosed within it thanks to their fiery constitution. Even if the surrounding environment and the other elements, notably water, play an essential role in the better constitution of those animals, here too Aristotle seems to attribute to vital heat a prominent role in the establishment of the hierarchy.

The link between relative excellence of animals and their hotter constitution is also attested in a passage of the *Parva Naturalia*. The passage is

[9] *GA* 2.1, 732b31–32. [10] *GA* 3.11, 761b13–15.

[11] *GA* 3.11, 762a18–24. For an attempt to situate Aristotle's theory of animal generation, including spontaneous generation, in the context of his theory of substantial generation, I refer the reader to Cerami 2015: 129–64. For more on the frothy bubble and its place in Aristotle's theory of animal generation, see Chapter 6 of this volume.

part of Aristotle's explanation of the existence of lungs in the animals that possess them. Aristotle asserts that the more valuable animals have a greater proportion of heat (τὰ τιμιώτερα τῶν ζῴων πλείονος τετύχηκε θερμότητος). He infers that it is for this reason that they also have lungs, since lungs ensure that the entire organism, including the other vital parts, is kept cool. He concludes that for the same reason those animals "must also have been assigned a more valuable soul (ἀνάγκη καὶ ψυχῆς τετυχηκέναι τιμιωτέρας)," that makes them "more valuable (τιμιώτερα) than the nature of plants." For a higher constitution always implies a higher nature.[12] In this case, the classification goes beyond the range of animals and bears on the entire domain of living beings.

At the end of *GA* 1, Aristotle extends the classification even more, so as to include also inanimate beings. In comparing living beings with inanimate beings, he tells us that we should consider plants to be more valuable than nonliving things, animals in general more valuable than plants, in that they all possess a degree of knowledge of a sort, and animals that use intellect yet more valuable. Again in *GA* 2.3, Aristotle ranks the capacity (δύναμις) possessed by all types of soul as more or less valuable depending on the kind of matter with which it is associated:

> Now, all soul-capacity seems to be associated with a body that is different and more divine than the so-called elements, and as the souls differ from each another in value and lack of value so too such a nature differs [trans. after Gotthelf and Falcon].[13]

This is also the reason why the human being, as the most valuable animal, stands upright on two feet. For the great amount of heat in his upper part orients him toward the upper region of the cosmos in which fire finds its own place.

In all these passages, heat, vital heat, and more generally hot matter are identified as the necessary constitutive element that makes living beings more or less valuable with regard to their constitution as well as their generation. If we leave the realm of animate beings, we realize that the same proportionality structures the order of the simple bodies. At the elemental level, fire – the element that is primarily characterized as hot – is repeatedly identified as superior to the other three elements. The relative excellence of fire is suggested notably in *GC* 1.3. In explaining the specific nature of substantial generation ("unqualified coming-to-be") as opposed

[12] *Resp.* 13, 477a11–25. Cf. *Resp.* 10, 476a6–15; 12, 476b13–19, especially 18–19.
[13] *GA* 2.3, 736b29–33.

to alteration and growth ("qualified coming-to-be"), Aristotle seems to admit that fire has a relative ontological superiority over the other elements. First, Aristotle defines fire as the product of an unqualified generation, contrary to earth, and associates it with "what is."[14] Afterward, he describes hot as a positive predication, and more of a form, and cold as privation; then, he suggests that fire, differing from earth according to this constitutive difference, is more of a substance and a "certain this" (τόδε τι) than earth.[15] The same idea is found in *GC* 2.8, where Aristotle explains that "fire alone – or more than all the rest – is akin to the form" (μόνον γάρ ἐστι καὶ μάλιστα τοῦ εἴδους τὸ πῦρ).[16]

All these passages are evidence of a relative superiority of heat and hot matter in Aristotle's cosmobiological framework: in the realm of living beings, a greater amount of heat is the necessary condition of the higher rank of certain species among others; furthermore, in the domain of simple bodies, heat, conceived as the elemental quality by which fire is primarily characterized, is the constitutive difference that explains the superiority of fire among the four elements. But do these assumptions entitle us to identify heat as such or vital heat as ultimately responsible for the excellence of more valuable animals and, by that token, to identify it as what provides the ultimate criterion of Aristotle's normative classification?

In a pioneering study, Gad Freudenthal argues that heat, identified as a "formative" agent and a fundamental feature of all natural beings, lies at the root of whatever persistence there is to be found in the world of generation and decay and gives rise to whatever "good" the sublunary world contains. For the same reason, he suggests that vital heat determines the entire *scala naturae* and guarantees the "continualist" vision of this scale, from the inanimate to the highest forms of life. Actually, Freudenthal further advocates that if the hierarchy were based on the different formal principles of natural beings and, more precisely, on their souls (vegetative, locomotive, perceptive, and imaginative), the scale of being would be discrete, rather than forming a seamless progression.[17]

In the pages to follow, I would like partially to resist this bottom-up explanation of Aristotle's hierarchical classification. I hasten to say that I do not want to challenge the idea that heat and, particularly, vital heat has a

[14] *GC* 1.3, 318b1–12. Aristotle suggests that the passage into fire is "unqualified" coming to be, whereas the coming to be of earth is "qualified" coming to be, though "unqualified" passing away (e.g., of fire). In so doing, and by appealing to the doctrine that he attributes to Parmenides, he concludes that, as far as generation and corruption are concerned, one could assume that fire and earth are as "what is" and "what is not."

[15] *GC* 1.3, 318b14–18. [16] *GC* 2.8, 335a14–20. [17] Freudenthal 1995: 65–70.

primary and necessary role in the development, structure, and functioning of living organisms. I will limit myself to suggesting that vital heat cannot provide the ultimate criterion of Aristotle's classification and the sole guarantee of its continuity. In order to understand the rationale of the scale, to use Aristotle's own words, we have to explain it on account of "what is better" and adopt "a principle from above."

In order to understand why vital heat is not the ultimate criterion of classification of animal species, we must first clarify the role that this corporeal principle plays in the phenomena related to the constitution and generation of living species and particularly in those animals that Aristotle identifies as the most valuable ones: the viviparous. The analysis of their generation and of the role played by the semen and the vital heat will show that the classification of species cannot be established merely on the amount or the purity of their heat. The gist of the argument I am about to offer is that heat cannot provide the ultimate criterion of classification because it is nothing but an instrument by which nature realizes what is good, identified as the internal end of living beings. In this sense, I would like to suggest that vital heat must be regarded as an instrumental cause in the realization of this end. For this reason, and in this limited sense, I would rather suggest that its presence in greater amount and in superior quality are a good for an animal only in a derivative way.

Reproduction and Instrumental Cause: The Derivative Good

The frequency of the references to vital heat in the biological treatises attests that the latter is a central explanatory concept in almost all the phenomena related to life. However, nowhere does Aristotle provide a systematic exposition of this notion, so that two questions remain partially unsettled. First, Aristotle does not explain clearly if or why the heat of elemental fire differs *essentially* from the vital heat proper to living beings. He gives some hints when he clarifies the nature and the role of the *pneuma* in the process of reproduction. There, Aristotle states that vital heat has a special affinity (κοινωνία) with soul and that it differs from the "so-called elements" in being more divine than they are.[18] Second, even if Aristotle focuses on the physiological aspect of vital heat's contribution in reproduction, he never defines clearly its specific causal role. In a great number of passages, he seems to suggest that vital heat is to be assimilated

[18] *GA* 2.3, 736b29–35.

to an efficient cause, but it remains difficult to understand in what way vital heat can be such a cause. Aristotle seems to explore two possibilities.

In examining the constitution of a new being, Aristotle suggests that heat should be seen as an efficient cause insofar as it is the principle of cohesion. He points out that the same principle also applies in the case of inanimate beings and the homoeomerous bodies that constitute living beings. This thesis is grounded in the study of the chemical properties of the four elements outlined in *GC* 2 and supplemented in *Mete.* 4. In both treatises, heat is identified as the main factor in explaining the phenomena that take place by concoction (πέψις). In *GC* 2, Aristotle explains that heat (as well as cold) is an active quality: as such, it acts on the passive qualities (wet and dry) and causes the concoction.[19] In *Mete.* 4, Aristotle defines concoction as what happens to everything when "the wet component is subjugated:" heat acts on the wet component, which serves as "indeterminate material" and it makes it more compact, dense, and dry.[20] Aristotle explains that all processes of concoction are characterized by the ability of the hot component to put together things of the same kind, to eliminate the residues and in so doing produce a homogeneous body.[21] From this point of view, the action of heat is essentially to associate bodies of the same kind.[22]

From a different but not conflicting point of view, Aristotle makes clear that the causal contribution of heat and, more properly, of vital heat in reproduction is to transmit movement. This is clear from the way in which Aristotle explains the phenomenon of procreation in the paradigmatic case of animals reproducing themselves by virtue of semen. In the study devoted to semen (*GA* 1.17–22), Aristotle first states that vital heat produces the concoction that transforms the ingested food into semen, as well as the concoction that the male semen activates in the female material. Thus, he points out that semen, being characterized by vital heat, is that from which (ἐξ οὗ) the new being comes to be as from a "first mover."

Aristotle makes clear what kind of mover semen is at the end of *GA* 1, when he defines semen as a residue of useful nutriment in its final stage[23] and, afterward, at the beginning of *GA* 2, when he describes the kind of movement that the semen triggers. Aristotle explains that semen

[19] *GC* 2.2, 329b24–31; cf. *Mete.* 4.1, 378b10–25. [20] *Mete.* 4.2, 379b32 ff.

[21] *GA* 1.18, 724b26 ff.

[22] In fact, the capacity to separate and remove residues is the indirect result of the combination of things of the same kind (cf. *GC* 2.1, 329b27 ff.).

[23] *GA* 1.18, 726a25–27.

incorporates an important amount of vital heat and that it is in virtue of this vital heat that it moves and is able to initiate movement, in that, as he says, 'heat is a motive force.'[24] He points out that semen is moved with the same movement as that in virtue of which the body increases (this increase being due to subdivision of the nutriment in its final stage). It is this same movement that semen communicates to the female residue when it enters the uterus and imparts form to the menses or καταμήνια.[25] It is in this way that the process of formation of the embryo is set in motion.

In this context, Aristotle explains that if the semen is a motive cause, it moves only insofar as it is a tool of the real and first agent of the generation, i.e., the male parent:

> [Semen] is no part of the embryo which is coming to be, just as nothing comes away from the carpenter to the matter of the timbers, nor is there any part of the art of carpentry in the thing which comes to be, but the shape and the form come to be present in the matter *from* that *through* the movement, and the soul (in which is the form) and the knowledge move the hands or some other such part in a certain sort of movement (different when that which comes to be from them is different, and the same when that which comes to be from them is the same); the hands move the tools, and the tools move the matter. Similarly, the male's nature too, in those which emit semen, uses the semen as a tool, one having movement in actuality, just as in the things which come to be by craft the tools are in movement; the reason is that the movement of the art is *in a way* in them [trans. after Gotthelf and Falcon, with modifications].[26]

In this passage, Aristotle states that the semen of the sire acts as an instrument in the process of fertilization. It is, though, clear that, more properly, it is the vital heat that fulfills this role. Actually, it is vital heat that carries the motion imparted by the sire. Indeed, it is thanks to the heat that the semen has the same movement that the nutritive soul possesses in actuality (ἐνεργείᾳ) and, again, it is thanks to the heat that it transmits the form of the male parent to the material principle provided by the female parent. Aristotle explains that, since the semen possesses the movement in actuality, the nature of the principle it contains is such that, once the motion has stopped, each part of the living being takes on its form and becomes animated. It is in this sense that we can conclude that the semen has only potentially the formal principle that it conveys, even if it possesses in actuality the movement that heat transmits.

[24] *GA* 2.1, 732a20. [25] *GA* 2.3, 737a18–22. [26] *GA* 1.22, 730b11–23.

For this reason, the heat of the semen is the cause that triggers the movement, but it is not, as such, the first principle of motion; it is nothing but an instrumental cause. The analogy with the products of craft shows that the vital heat of the seed has the same role that a tool plays in what is produced by craft. According to this analogy, the male parent, his soul, his sexual organ, his semen endowed with the vital heat are, in the formation of the embryo, what the craftsman, his skill, his hands, and his tools are in craft production. Semen, by its vital heat, is, in the process of fertilization, what a tool is in the production of an artefact.

The same analogy is used in *GA* 2 to clarify the link between vital heat, the generative movement, and its principle. Here, however, the role of instrumental cause is explicitly attributed to vital heat and not to semen:

> Just as the things which come to be by the agency of the art come to be by means of tools – it is truer in fact to say "by means of the *movement* of them" (this [movement] is the actuality of the craft and the craft is the form of that which comes to be in another) – so does the nutritive soul-capacity. And just as in the animals and plants themselves later on it produces growth from the nutriment using heat and cold as tools (for the movement of the soul-capacity is *in* these, and each thing comes to be according to some ratio [λόγος]), so also at the very beginning it constitutes the thing that comes to be by nature [trans. after Gotthelf and Falcon].[27]

In this case too, Aristotle makes clear that the products of a skilled craftsman and the products of animal generation have something in common. Just as we would not say, about an axe or anything else made by a blacksmith, that it is the product of fire alone, so also when it comes to a living body or any of its parts, we cannot state that it is vital heat alone that generates them. Fire and water can make iron harder or softer; but what makes a sword from iron is the carefully judged movement imparted by the craftsman to his tools in the exercise of his skill. In an analogous way, hot and cold can at most produce in a given body properties such as hardness, softness, elasticity, friability; but they cannot explain the fact that at some point in the process of generation one part becomes an organ and another flesh. This is due to the movement proceeding from the sire and, more precisely, to the proportion that characterizes it.

Thus the same holds for craft production as for animal generation: as the craft is the form, while the fire is a tool in the production of the craftsman, in a similar way the nutritive soul is the form, while the vital heat is an instrument in the accomplishment of its activity. The nutritive

[27] *GA* 2.4, 740b25–34.

soul provides the movement with its "ratio" (λόγος), which is to say its proportion and measure; for the same reason, it is the very first principle of its release and it is also what uses heat and cold as instruments in order to transmit the movement that characterizes its activity.

The instrumental character of the action of heat is also brought out in the *PA*, where Aristotle also confirms that in the constitution of the body heat can be considered the first substrate of the soul:

> While some crudely posit fire or some such potential to be the animal's soul, it is perhaps better to say that soul is constituted in some such body. This is because among bodies the hot is the one most able to assist with the functions of the soul; for nourishing and producing change are functions of soul, and these things come about most of all through this potential. Saying fire is the soul, then, is like saying that saw or auger is the carpenter or carpentry because the function is accomplished when they are near each other. That animals must partake of heat is clear from these considerations [trans. after Lennox][28]

The soul is neither fire nor heat. However heat maintains a special relationship with soul that allows us to define it, from the point of view of the constitution of the animal, as the *substrate* of the soul and, from the point of view of the generation and, more generally, of the nutritive functions, as an *instrumental cause*. Even if Aristotle seems to attribute a logical priority to the capacity of being an instrument, being the substrate of the soul and being its instrument in the realization of its functions are the two aspects of a single phenomenon: being a substrate is what characterizes heat from a static point of view, being an instrument is what characterizes it from a dynamic point of view.

Therefore, heat constitutes the substrate in which soul exists, since it is the most appropriate and effective instrument for soul to use in exercising the functions that are proper to it, particularly as they relate to the transmission of psychic movement in the processes of generation and of growth. In itself, vital heat can be neither the principle of identification of a living body nor the principal cause of its generation. It is the necessary instrument for the realization of soul's functions, as the saw or the auger is the necessary instrument for the production of a piece of carpentry. However, precisely because vital heat is nothing but an instrument, it can determine neither the generation of the animal nor its growth. As Aristotle states, the reason for this is that the movement conveyed by vital heat must have a proportion, a λόγος, and that this λόγος can be provided

[28] *PA* 2.7, 652b7–16. Cf. *DA* 1.2, 403b31

neither by the heat nor by the cold. On the basis of what Aristotle says in his treatise *On the Soul*, we can conclude that heat, as such, cannot set a limit to its own action, since, like fire, its growth extends indefinitely as long as there is combustible available. Indeed, all biological phenomena, and notably concoction, are essentially characterized by a limit, for all systems that nature forms have an order of size and of growth. But this, insists Aristotle, cannot but be the result of the soul.[29] According to this explanatory paradigm, thus, we must conclude that for Aristotle to state that the coming to be of an animal is caused by heat is as absurd as to state that Aphrodite's statue has been sculpted by Praxiteles' chisel.

Although the process of generation and growth cannot occur without heat, from a quantitative point of view, heat cannot determine the size of the generated animal and more generally any measure that characterizes it. For the same reason, from a qualitative point of view, we must admit that heat cannot be the ultimate criterion of the superiority of one species over another in the animal hierarchy. In fact, if heat is nothing but the necessary instrument of the organizing principle of the living being, it cannot *as such* determine the quality and excellence of one species as compared with another. For, what defines the best is also what defines the good. Now, the instrument cannot define what is good, even if it is a necessary condition in bringing it about. As a necessary but not a sufficient condition, the instrument could only be a good in a derivative sense. It follows therefore that the amount of heat can be no more than a secondary criterion of excellence in animal ranking. In fact, properly speaking, only the *act* for the sake of which the instrument is used can provide such a criterion. But what act, one must ask, is this?

Good, Better, and Self-Preservation: To Assimilate the Other to Persist in Actuality

The foregoing line of reasoning might lead to the conclusion that since the criterion of animal ranking is not the amount of heat, because heat is only a necessary instrument, the nature of the agent principle governing the instrument, namely the soul of the living being, provides such a criterion. In other words, one could assume that the animal hierarchy is established

[29] *DA* 2.4, 416a9–15: "On the other hand, some people think, that the nature of fire is quite simply the cause of nutrition and growth [...]. Though it is, in a sense, the auxiliary cause, it is not a cause without qualification: it is soul that is that; for fire grows indefinitely so long as there is fuel for it, but all naturally constituted things have a limit to their size and growth, a ratio (λόγος) which they preserve, and these belong to soul, not to fire, to ratio, not to matter" [trans. after Ross].

on the basis of the different types of souls *as such*: among living things, the species with the more valuable forms of soul would be superior to the others, as throughout animate nature as a whole, rational souls would rank above animal souls and animal souls above vegetative. On this interpretation, therefore, one life-form would be more valuable than another if it had all the capacities possessed by the other *plus* some further capacity.

This hypothesis, however, encompasses at least two difficulties that call into question its very relevance: first, the ontological autonomy of each living species and the fact that each substantial form is an end and as such a good for the species itself hinder the establishment of a comparative scale; second, as we have already observed, even if we posit such a scale, this hypothesis cannot guarantee its continuous nature. Indeed, substantial forms *as such* cannot provide a unitary principle for a unitary scale. We have to find a truly common principle in order to establish a truly unitary scale.

In order to throw light on the criterion that *can* provide Aristotle's normative classification, I will take advantage of another of Allan Gotthelf's groundbreaking papers.[30] Aiming at showing the dependence of goodness on natural ends, Gotthelf points out that in all things that have a function, the good and the well reside in the function, defined as a certain kind of life. To perform a function is to achieve the end. To perform it well is to achieve that end fully; to perform it less than well is to fall short of the end in some way. From this assumption, Gotthelf argues that the goodness of a state or process of an organism lies in its contribution to the life, to the preservation and realization of the being of that organism, where the concept of life is not itself defined in terms of a prior concept of the good.

In Gotthelf's analysis, the notion of good at stake, as well as the natural end that defines it, are proper to each animal species. The good is what preserves the species, the best is what makes the preservation of this particular kind of life easier, steadier, or more durable, by enhancing the performance of the function(s) that makes it possible.[31] In other words, what is good and what is better are always relative to a given kind of life.[32]

In this sense, the framework suggested by Gotthelf may seem to be of little use when comparing types of life and not relative contributions to a

[30] Gotthelf 1988 (reprinted in Gotthelf 2012a: 45–66).

[31] In the words of *De long.* 4, 469a28–31, one can safely say that the measure of the best is that the result "most enables each of the parts fully to perform its proper function."

[32] In this sense preservation is always *essentially* preservation of a distinctive being and a distinctive kind of life. See *DA* 2.4, 416b14 and the surrounding context.

given type of life. To counter this possible objection, I would like to suggest that the same notion of good as an internal end is at the foundation of Aristotle's scale of being, even if in a more indirect way. It is not the instrument that sets the scale of perfection, but the goal, namely the function that the instrument enables to realize.

On this view, the good that provides the criterion for the scale is the function that all the members of the given class share. Accordingly, the most valuable member is the one that has a more powerful version of the corresponding capacity. As in the case of what is good as a relative contribution to a relative end, what is better within different ends, to quote Allan Gotthelf, does not involve "the importation of any criterion of value other than contribution to life."[33] In what follows, I will support this claim and try to establish the extent of its validity.

It is crucial to remark that in ranking different kinds of life as in ranking different contributions to a given kind of life, the choice of the function determines the rationale of the scale. On this principle, Aristotle can set up different classifications according to the kind of being in question. However, as we will shortly see, it is still possible to identify a single classificatory criterion to use as the basis of a single continuous scale of being, even if the single criterion involved and the single scale constituted according to it are only such by analogy.

I will consider first the ranking of animals that we find in the *GA*. As we have seen, this classification takes in, without distinction, all entities defined as living beings. According to the suggested hypothesis, we need to identify their common function in order to understand the ranking. Aristotle explains that the function (ἔργον) of every living being perfect in its nature, whether animal or plant, is to generate another individual similar to itself in form and thereby to participate in what is divine and eternal:

> In fact, whether plant or animal, this is present in all alike, namely the nutritive capacity. And this is what has the power to generate another like itself. The reason is that this is the function of everything naturally complete, both animal and plant [trans. after Gotthelf and Falcon].[34]

The reproduction of another like itself is the common function that all living beings share, that is, animals as well as plants. This function defines living beings *qua* living beings and constitutes their internal end and as such their good. Therefore, this primary function is the common principle

[33] Gotthelf 2012a: 57. [34] *GA* 2.1, 735a18–19.

at the foundation of the unitary scale. And since it is the formal principle common to all living creatures that determines the function, we can conclude that it is its formal *power* that provides the ultimate criterion of the ranking: it is the nutritive power that explains the existence of the scale of perfection within living beings.[35]

The ultimate standard of perfection for a species is not the possession of a greater amount of heat but rather the capacity of each of its members to generate another being like itself and, in so doing, to convey its own form and assimilate matter in the most direct and effective way. The most perfect animals are those that by their nature are most able to perform this function. All of them are capable of achieving this end, but not in the same way. All of them can do it, but some do it better than others. For this reason, live-bearing animals surpass all other animals in excellence, while the egg-laying animals rank second in the scale. For live-bearing animals are better at fulfilling this function, in that they give birth in a more *direct* and independent way to individuals like themselves perfect in respect of quality. It becomes then clear why the scale of animals *qua* living beings is established according to their different modes of reproduction:

> We should realize how well and in an orderly sequence nature renders the coming to be. The more complete and hotter animals render their young complete with respect to their quality (but with respect to their size in general none of the animals do, for all take on growth after having come into being), and these animals indeed generate inside themselves right away. The second in line do not produce complete animals within themselves right away (since they produce live young after having first produced eggs), but they produce live young externally. Others do not generate a complete animal but generate an egg, and this egg is complete. Others, having a nature still colder than these animals, generate an egg though not a complete one, but it gets completed outside as in the case of the kind constituted by the scaly fishes and the soft-shelled animals and the soft-bodied animals. The fifth and coldest kind, does not produce eggs out of itself, but this sort of condition occurs in this case externally, as was said: for the insects produce grubs in the first place, but the grub, having developed, becomes egg-like (for the so-called chrysalis has the potentiality of an egg), then from this thing there comes to be an animal in the third change, reaching the end of the coming to be [trans. after Gotthelf and Falcon].[36]

[35] Aristotle concludes that, for the same reason, the nutritive principle must be the first to be engendered, and that it must be the same soul-principle that accounts for the process of growth, which is seen as sharing the same goal: generating another individual similar to the first.

[36] *GA* 2.1, 733a32–733b23.

The more stages the ensouled matter has to go through in order to become similar to the sire, the less perfect the animal is. From this point of view, the live-bearing male that generates another male is the absolutely most perfect animal among living beings, since it is capable of reproducing an individual which is similar to itself in the most direct way. The live-bearing are the most perfect animals, because in virtue of a superabundance of heat they produce young which are perfect "in respect of quality," that is, offspring that are already fully alive when separated from their mother. Conversely, it follows that a being is less perfect if its capacity to produce another individual like itself is less efficient owing to its colder nature.

The material criterion is, so to speak, the visible side of animal classification; it is the relative excellence of the soul of the living being, or rather its more or less powerful capacity to act on a material and to produce another individual of the same species, that determines the superiority of a given species. Hottest animals are the most valuable animals insofar as heat is the physical agent that better conveys the form and acts on matter so as to make it similar to the generating principle or, as Aristotle says in *GA* 4, to "bring it to itself."[37] The normative criterion is therefore the capacity to transmit its own form and in so doing to assimilate what is different.

On this yardstick, one could integrate in the ranking not only egg-laying and other animal groups, but also plants which, because of their cold and earthy nature, give birth to a seed and are, for the same reason, the form of life furthest remote from the perfection that characterizes the highest order of living beings.

According to this reading, we can also better understand in what sense Aristotle suggests that the explanation of the differentiation of the two sexes depends on "a principle taken from above."[38] This explanation is provided by the capacity common to all living beings, namely the ability to assimilate what is different in order to ensure their own survival, as far as that is possible for generable and corruptible beings. By having a more powerful capacity to assimilate what is different and hence for survival, the male principle is more valuable than the female one, which is more properly what is assimilated and does not survive by itself. This same idea therefore explains why Aristotle denies that living beings unable to reproduce themselves (including, therefore, children and sterile individuals) are substances in the fullest sense.

From this point of view, motion and sense-perception, which are not common to all living beings but proper to animals, provide a

[37] *GA* 4.1, 766b15–18. [38] *GA* 2.1, 731b23: ἄνωθεν ἔχει τὴν ἀρχήν.

complementary but not different criterion, in that both capacities improve the chances of self-preservation.[39] From this point of view, indeed, it is clear that the criterion of self-sufficiency introduced in *GA* 2.1, 732a11–23, and determined by the capacity to move stems from the criterion of self-preservation and not *vice versa*.

In all the cases considered, the capacity to reproduce another like itself provides the criterion according to which some living beings are more valuable than others. But how far can this explanation be extended beyond the case of animals? Is it possible, on the basis of the same criterion, to extend the classification to other beings, notably inanimate beings? To answer this question, it is crucial to remember that reproduction is for Aristotle an extension of the capacity for self-preservation.[40] The animal is unable to preserve itself eternally, so it does the next best thing toward its preservation: it leaves behind something essentially like itself. For Aristotle, the aim is self-preservation, not preservation of the species. The latter is a consequence of the former.[41] Thus, according to the same hypothesis, but on a more general perspective, we can conclude that the most valuable form is the form that is most able to assimilate what is other than itself in order to persist, that is, to preserve itself.

This same principle can be taken back and applied by analogy to the case of the elements: the classification of the four elements according to a scale of perfection is dictated by the fact that an element is more or less capable than another of imposing its form and thus assimilating that on which it acts. It is in virtue of its being the most active element that fire is also the most perfect. Put differently, it is not because fire goes up that it is more perfect, but rather because the quality that characterizes it, namely heat, makes it more capable of acting on the elements that are more passive by their nature.[42] As Aristotle says in *Mete.* 4.1, 379a1, heat generates by "ruling" matter.

We can gather that one consequence of this is that fire's perfection flows back on entities in which it is present in proportion to the quantity they have of it, so that animals that possess a greater amount of heat will be more perfect than animals that have less. But this is not to say that the simple possession of a certain quantity of fire, as such, explains the degree

[39] On the role of sense-perception in self-preservation, see Rashed 2002.

[40] *DA* 2.4, 415a23–b7; *GC* 2.10, 336b25–34; and *GA* 2.1, 731b18–732a1.

[41] Gotthelf 2012a: 129. Cf. Rashed 2002.

[42] Hence, the location of the element is not to be taken as the criterion of relative excellence. It is rather the other way round: the location is more or less valuable because of the element that is found there.

of perfection attained by a composite being; what determines that is the natural ability of the heat it brings with it to dominate the more passive components, a function that corresponds analogously to the function that makes an animal what it is, *viz.* its ability to act on what is passive in order to impose its form and thus ensure its survival.

The criterion of self-preservation explains why live-bearing animals are the most valuable living beings, and why fire is more valuable element. But, again, is it possible to extend the principle even more so as to include in the ranking everything that exists? To answer this question we must look more closely at what is meant by "self-preservation." If the perfection of the individual lies in its proper activity, that is, the *telos* or end that constitutes the good toward which the organism's function is directed, then the notion of actuality is, in a sense, prior to that of self-preservation. In other words, actions aimed at self-preservation are, as such, attempts by an organism to secure its persistence in actuality in the fullest sense.

Thus, by going a step further, we can conclude that an organism's capacity to preserve itself is, as such, a capacity to persist in actuality. In so doing, therefore, we can further infer that this capacity can be placed at the foundation of a wide-ranging scale of being. In fact, the capacity to persist in actuality, proper not only to animate and inanimate perishable beings, but more fundamentally to all individuals in the Aristotelian cosmos, is the criterion upon which everything that exists can be ranked on a single scale, which is unique scale of being. This is what explains, I would finally suggest, the legitimacy of the normative classification advanced at the outset of *GA* 2:

> Since of the things that exist some are eternal and divine, while others admit of being and of not being, and the fine and the divine are by their own nature always cause of the better in things which admit of it, and the non-eternal is something which admits of both being and not being and of sharing in both the worse and the better, and since soul is better than body, and the ensouled is better than the soulless because of soul, and being is better than not being, and living is better than not living – due to these causes, there is the coming to be of animals. Since it is in fact impossible for the nature of such a kind to be eternal, what comes to be is eternal according to the way in which it admits of it. Thus, it cannot be eternal in number (the reason is that the substance of the things that exist is in the individual, and it would be of such a sort if indeed it was eternal), but it can be eternal in form. That is why there exists always a kind: of human beings, animals, and plants. Since the female and the male are their source, it would be for the sake of the coming to be that there are the female and the male in those that have them. But since the proximate moving cause, in which are

> found the definition and the form, is better and more divine than the matter, it is better also for the better thing to be separated from the worse. For this reason, in those animals in which it is possible, and to the extent possible, the male is separated from the female, given that the source of movement (which is where the male is found in things which came into being) is better and more divine, whereas the female is matter. And the male comes together and mingles with the female for the activity of coming to be because this is common to both [transl. after Gotthelf and Falcon].[43]

By extending the criterion of self-preservation as suggested above, we can arrive at a better understanding of Aristotle's scale of perfection and the nature of its universality: something is more or less perfect according to the degree to which its being is closely related to actuality in the fullest sense. In the realm of sublunary beings, the capacity of persisting in actuality is necessarily linked to the capacity of assimilating what is other than itself. The same principle however can be extended to all beings of the Aristotelian cosmos, not only in the case of beings subject to generation and corruption, but also at the level of the world of the incorruptible beings.

Indeed, it should be stressed that in the realm of "the being that is always the same," the criterion of perfection is based on the capacity of persisting in actuality *as such*, whereas in the case of beings subject to generation and corruption, it is based on the capacity of assimilating what is different and in so doing reproducing something similar to themselves. Still, we can conclude that this capacity provides us with the criterion of a single ontological scale that encompasses all the beings of the Aristotelian world. On this perspective, the unmoved mover, as pure actuality, is the most excellent being, followed by the heavenly bodies which in their turn are characterized by a single eternal and continuous activity; live-bearing animals occupy the third place, not only because they perpetuate themselves at the level of the species, but also because they can lay claim to a degree of personal immortality by directly reproducing a quasi-replica of themselves.

Conclusion

It is because self-preservation is one aspect of a universal principle that applies by analogy to all beings in the Aristotelian cosmos that we are justified in concluding that being is more valuable than nonbeing, movement is more valuable than rest, life more valuable than privation of life,

[43] *GA* 2.1, 731b24–732a9.

and intellectual soul more valuable than sensitive soul. The most perfect being is that which is least alloyed with what is not actuality and which is closest to pure actuality. On this basis, we can safely conclude that the principle at the foundation of the scale of perfection is neither the amount of heat or its quality, nor strictly speaking the substantial form of the different classes in the ranking. Still, there is a unitary classification even if its unity is not by univocity but by analogy. In the realm of generation and corruption, the analogy is established on the function common to all generable and corruptible bodies: the capacity to assimilate what is different in order to survive. At the most general level, the criterion on which the scale of perfection is based is nothing other than the capacity to persist in actuality: actuality is the furthest remote principle, which provides the entire cosmos with its stability and its structure. But here we enter the domain of another science: the science of being *qua* being.

PART III

Hybrids, Male and Female, Particular Forms, and Monsters

CHAPTER 8

Hybridity and Sterility in Aristotle's Generation of Animals

Jocelyn Groisard

Introduction

The biological turn in Aristotelian studies has led to growing doubts about Aristotle's so-called essentialism. Aristotle's biology has been less and less conceived of as a classification of fixed species corresponding to eternal forms. The notion of "species" as the core of Aristotelian biology has been criticized in several ways. One criticism emphasizes that the focus in Aristotle's biology is on *parts* of animals and their functional analogies between different species, rather than on animals as wholes, i.e., as organized sets of parts to be classified in species according to their distinct characteristics.[1] Another line of criticism has cast doubts on the generally accepted notion that Aristotelian species are eternal.[2] For example, if specific eternity is just the transmission of the male form from parent to offspring, such cases as spontaneous generation do not seem to fit easily into a system of eternal specific forms being transmitted or reproduced again and again without change.

One issue sometimes mentioned by critics of a fixist interpretation of Aristotle's biology is the possibility, which he admits of in *GA* 2.7–8, of hybridity.[3] Since hybrids are born to parents of different species and do not belong to any of these species, hybridity creates new living forms in a process of diversification which in itself seems to contradict the fixist biology usually assigned to Aristotle. Far from constituting a closed system of fixed species corresponding to eternal forms or essences, Aristotelian living forms could then be seen as both heterogeneous in the way they are generated and dynamic according to the possibility of hybridity. As opposed to the traditional view of Aristotelian biology as a kind of archaic taxonomy classifying species devoid of any evolution, one could even imagine it as implying an open set of organic forms that would be able

[1] Pellegrin 1990: 37–48. [2] Lennox 2001: 131–59. [3] Balme 1992: 97.

to diversify according to the conditions of spontaneous generation or by successive hybridization.

Yet leaping to such a conclusion would be wrong. First, it is obviously anachronistic to introduce a kind of evolution in Aristotle's biology, since nowhere does he say anything that could suggest a continuous evolution of species as a result of successive hybridization. Second, the best part of Aristotle's explanations about hybridity in *GA* 2.7–8 aims at defining conditions that strictly limit the possibility of generation between animals of different species. Aristotle mentions two kinds of limitations, one being that hybridity is possible only between species possessing a strong similarity, the other that the offspring may be sterile, as can be seen in the case of mules. One might think that, taken together, these limitations would be seen as blocking an indefinite diversification of species through hybridity, but in fact this is *not* the case: only the necessary similarity between parents' species works as a general limiting condition of hybridity, whereas the sterility of mules is considered by Aristotle as an exception, given that other hybrids are usually able to reproduce according to him.

This is why this paper, which focuses on *GA* 2.7–8, will consider separately, first, the general conditions of hybridity according to Aristotle,[4] then the special case of mules' sterility, about which he adduces, and refutes, earlier theories by notably Empedocles and Democritus.

Conditions of Hybridity

Aristotle's very first statement about hybridity in *GA* 2.7 is that "mating happens according to nature between animals of the same genus,"[5] which implies that hybridity is somehow against nature. While being possible,

[4] Only animal hybridity is considered here. We have no text by Aristotle about hybridity in plants. In his *Historia plantarum*, Aristotle's disciple Theophrastus compares grafting to sexual intercourse (2.1.4.11–12) and announces more detailed explanations that can be found in his *De causis plantarum*: there, however, Theophrastus does not return to the comparison with sexual reproduction and minimizes the difference between grafting and other cultivation techniques by saying that the scion uses the rootstock just as if it were earth (1.6.1.3); furthermore, Theophrastus focuses on grafting between similar plants, or at most of a wild variety on a cultivar, for example a wild olive-tree on a cultivated one (1.6.10). The *De plantis* by first-century BC Peripatetic philosopher Nicolaus of Damascus makes the same comparison between grafting on a rootstock and planting in soil, then distinguishes several cases depending on whether scion and stock are of same species, similar species or different species, with some examples (apple tree on apple tree, apple tree on pear tree, ivy on all kinds of trees) but no further comments about hybridization. In sum, in the Aristotelian tradition, grafting in plants and hybridity in animals shed little or no light on each other; this is of no surprise, for Theophrastus himself at the beginning of the *Historia plantarum* warns against making undue comparisons between plants and animals (1.1.4.1–2).

[5] Aristotle, *GA* 2.7, 746a29–30: "γίνεται δ' ὁ συνδυασμὸς τοῖς ζῴοις κατὰ φύσιν μὲν τοῖς ὁμογενέσιν."

hybridity is just an exception to a general rule of generation according to which animals of different species do not and cannot mate. Aristotle immediately states the condition of this exception: mating is possible "even between animals having a close nature although not devoid of difference according to species."[6] This condition seems straightforward, but it is also rather vague and raises some problems.

The first issue is how to distinguish between natural generation and hybridity. If hybrids are born to parents which have some difference according to "species/form" (τῷ εἴδει), conversely natural generation should be between animals of the same εἶδος. But Aristotle has just said that natural generation happens between "animals of the same *genus*" (τοῖς ζῴοις τοῖς ὁμογενέσιν). One could easily ignore this overlapping between species and genus, since it is well-known that in Aristotle these terms do not refer to determinate levels in a classification and can often be substituted to each other; another term later used by Aristotle, ὁμόφυλα (746b 9: "of the same family"), clearly shows that terminological consistency is just not at stake here: for Aristotle, all three terms "species" or "genus" or "family" can refer to a group of animals inside which generation is considered natural and it is only according to an anachronistic convention, rather than Aristotelian usage, that we speak of hybridity as generation between different *species*.

Yet this inconsistency in terminology has a more theoretical reason, which is the lack in Aristotle's biology of *species* as a fixed taxonomical level inside which generation would be natural and above which hybridity would begin. The boundary between natural generation and hybridity is not clear and this appears in Aristotle's examples, which are sometimes ambiguous: when he says hybrids can be born "in the case of dogs, foxes, and wolves,"[7] does he mean hybridity between, e.g., a dog and a wolf or fox,[8] or between, e.g., dogs of different races?[9] Here the first reading is certainly better, because dogs, wolves, and foxes have at the same time a combination of similarity and difference which seems to meet the condition of hybridity as happening between animals of close nature but

[6] *GA* 2.7, 746a30–31: "οὐ μὴν ἀλλὰ καὶ τοῖς μὲν σύνεγγυς τὴν φύσιν ἔχουσιν, οὐκ ἀδιαφόροις δὲ τῷ εἴδει." See Cerami 2015: 159 for a detailed discussion of this text.

[7] *GA* 2.7, 746a33–34.

[8] In *GA* 2.4, 738b31–32, Aristotle refers to "hybrids born of a dog and a fox"; in *HA* 7.28, 607a1–3, he mentions cases of hybridity between wolf and dog as well as fox and dog.

[9] In *HA* 8.1, 608a31–33, Molossian and Laconian dogs are said to give birth together to dogs of exceptional resistance, but Aristotle's formulation does not make it clear if he considers this a case of hybridity.

different in species. But one cannot completely rule out the second reading, because in some cases the same name can apply to animals different enough to give birth to hybrids: for example, when Aristotle says that "hawks differing in species seem to mate with each other,"[10] he implies that animals similar enough to be called by the same name "hawks" can also be different enough to give birth to hybrids.

Just as vague as the limitations between natural generation and hybridity is the condition making this latter possible rather than impossible: Aristotle says hybridity is possible between animals of "close nature," but what does this precisely mean?

A first hypothesis would be to take as criterion of this closeness in nature a general similarity, such as between dogs, wolves, and foxes. Several of Aristotle's examples seem to confirm this interpretation: "Indian dogs" are born of a dog and a "wild beast similar to a dog;"[11] and hybridity is possible between birds sharing some common characteristics, e.g., between partridge and fowl, which are both "salacious" birds;[12] or between hawks of different species, all of them with crooked talons.[13] These common characteristics have a taxonomic quality since they define larger groups of animals between which generation is possible thanks to their similarity even if they have some difference in form: for example, hawks of different species are included in a larger group of "hawks" and all hawks are included with other birds of prey in an even larger group of birds having crooked talons. Yet this is not a systematic taxonomy since the boundaries of these larger groups remain rather vague and they are based on very heterogeneous criteria such as sexual behavior for "salacious" birds, morphology of one part of the body for birds with crooked talons, or, for hawks, a general resemblance allowing the use of the same denomination for all of them.

Even if the taxonomical level of hybridity remains a rather vague margin between natural generation and impossible hybridization, Aristotle adds two conditions that seem more precise: two animals of different species but close in nature can mate if (1) their size is similar and (2) the length of gestation is equal in both species.[14] Aristotle gives no further explanations about these conditions, but they are not so difficult to understand.

[10] *GA* 2.7, 746b2–3.

[11] *GA* 2.7, 746a34–35. In *HA*, this "wild beast similar to a dog" is called τίγρις (7.28, 607a3–4).

[12] *GA* 2.7, 746a35–b2.

[13] *GA* 2.7, 746b2–3.

[14] *GA* 2.7, 746a31–32; cf. *GA* 2.4, 738b28–30, and *HA* 8.28, 606b20–23.

Condition (1) can be understood either at the level of [a] reproductive parts or [b] of male semen and female menses.

[a] One could imagine two species very similar in form, but of very different size, such as a cat and a tiger, so that male and female sexual organs could not fit together in copulation or the uterus would be too big or too small for proper embryo development. So similarity in size is clearly an additional condition to close nature.

[b] Since for Aristotle fecundation happens when the male semen cooks the female menses by its inner heat, there must be a kind of proportion between them so that the right cooking process occurs.[15] But this proportion has a quantitative bearing, as is clearly shown by Aristotle's explanations in *GA* 4.4 about the difference between multiparous and uniparous animals: in uniparous species, the quantity of the female's menses is just enough to form one embryo, whereas in multiparous species menses are several times the necessary quantity, so that they can be divided into several embryos; accordingly, on the male's side, semen quantity must be sufficient to cook one or several embryos depending on the species being uniparous or multiparous, but always in the right proportion, because if too abundant it would overcook the menses and make them dry.[16] Aristotle concludes that "there must be some proportion (συμμετρία) between the female residues and those from the male,"[17] the general principle behind this condition being that "it is a given quantity that can operate upon a given quantity."[18] So for hybridity to be possible, there should be the right ratio between the semen quantity in the male's species and that of menses in the female's one, and this proportion is more likely to obtain between species close in size.

Condition (2) about the length of gestation is closely related to the previous one, since for Aristotle the time needed for the development of the fetus depends on the size of the full-grown animal. This issue is dealt

[15] *GA* 4.2, 767a15–23. On the importance of proportion (συμμετρία) and the various processes by which a lack of proportion leads to the generation of females, or children that do not look like their parents, or even monsters, see Lefebvre 2014: 207–72.

[16] *GA* 4.4, 771b33–772a22. [17] *GA* 4.4, 772a17–19.

[18] *GA* 4.4, 772a32: "ἐκ ποσοῦ τινος ποσόν τι τὸ ἐργαζόμενόν ἐστιν." Here ἐκ ποσοῦ τινος refers to a given quantity of female menses from which the action of a given quantity of male semen produces the embryo.

with in *GA* 4.10, where Aristotle considers whether gestation time is proportional to lifespan or to size. The example of the horse, which has a shorter lifespan than man but a longer gestation time, is used to opt for the latter criterion.[19] In the case of hybridity, an equal gestation time in the parents' species is necessary to allow sufficient time for the development of the offspring,[20] the size of which as an adult will be similar to that of its parents.

Although Aristotle does not state in great details the conditions of hybridity, it is clear that for him it is limited by a set of restrictions including closeness in nature in spite of some formal difference, similarity in size, and equal gestation times – as well as more incidental conditions such as coexistence in the same environment.[21] Hybridity, therefore, remains a marginal phenomenon, and in no way could it lead to a continuous diversification of species, challenging the fixist frame of Aristotelian biology. Moreover, since Aristotle's comments on hybridity in *GA* 2.7–8 are mainly concerned with the issue of mules' sterility, one could expect him to use the lack of fertility of crossbred offspring as a further limit of this marginal type of generation. Yet, as already said above, this is not the case.

Mules' Sterility

Just as hybridity is an exception within generation, so mules' sterility is considered by Aristotle as an exception within hybridity: "although all animals born from such a union are themselves seen to mate together, and to be able by uniting to give birth to males as well as females, mules alone are sterile among such animals; for they cannot give birth either between themselves or by union with others."[22]

[19] *GA* 4.10, 777a31–b16.

[20] In *HA* 6.23, 577b7–8, Aristotle states that, in the crossbreeding of horse and donkey, the prevailing gestation time is that of the male's species. One can easily understand why, since the male gives the form, whose development during pregnancy requires a specific time. In practice this principle is of little consequence since gestation time is supposed to be equal for crossbreeding to be possible (eleven months for horse and donkey according to *HA* 6.22, 575b26–7, and 6.23, 577a24).

[21] Both *GA* and *HA* report on an apparently common belief that there are many hybrids in Libya because scarcity of water makes animals of different species gather around springs (*GA* 2.7, 746b7–11; *HA* 7.28, 606b18–21). In *HA*, Aristotle adds that the need to drink also tames them (606b23–24), which helps them to cross the species barrier by reducing mutual distrust.

[22] *GA* 2.7, 746b12–16: "τὰ μὲν οὖν ἄλλα τῶν ἐκ τοιαύτης μίξεως γινομένων συνδυαζόμενα φαίνεται πάλιν ἀλλήλοις καὶ μιγνύμενα καὶ δυνάμενα τό τε θῆλυ καὶ τὸ ἄρρεν γεννᾶν, οἱ δ᾽ ὀρεῖς ἄγονοι μόνοι τῶν τοιούτων· οὔτε γὰρ ἐξ ἀλλήλων οὔτ᾽ ἄλλοις μιγνύμενοι γεννῶσιν".

Whereas it is clear that for Aristotle hybrids are able to reproduce, this can be understood in two ways: does it mean that animals born of some given hybridization are able to reproduce *between them*, or that they can themselves give birth to second-level hybrids by intercourse with other species? This question has an important bearing on the degree of species diversification allowed by hybridity: does it just create viable species, within each of which hybrids are able to reproduce together, or can it create a string of new species indefinitely diversifying by successive hybridization with other species? For mules, Aristotle explicitly says that they cannot reproduce either between themselves or with other animals, but for hybrids in general he does not distinguish between these two cases. In the above quotation, one could take the variation between συνδυαζόμενα πάλιν ἀλλήλοις and μιγνύμενα to mean such a distinction between, respectively, copulation within any hybrid species and a new "mixture," i.e., a second-level hybridization; and to argue that μιγνύμενα here refers to hybridity, one could use the occurrence of the related noun μίξεως at the beginning of the same sentence by giving it the meaning of "mixture (of species)" as some translators actually do.[23] Yet in the context of *GA* μίξις and μίγνυσθαι just refer to copulation, so the most probable interpretation here is to take both συνδυαζόμενα and μιγνύμενα as meaning the same thing, i.e., the reproduction of hybrids between them (ἀλλήλοις). Aristotle's text does not rule out second-level hybridization, but makes no hint at it, so it seems more cautious to take his general rule of hybrids' fertility as meaning closed reproduction within a given crossbred species.

Mules are an exception to this rule, and also within infertile animals. Before discussing this case, Aristotle makes a long digression to explain that sterility is usually an individual deficiency due to the normal nature of the species being somehow altered: sterility can happen because of a congenital deformity of sexual parts, or because residues that should form semen are diverted into the accumulation of fat, or because ducts conveying these secretions to sexual organs are not functioning properly.[24] But these accidental causes for sterility do not apply to mules since, as Aristotle says, "in men and other genera such an infirmity happens at the level of particulars, whereas in mules the genus as a whole is sterile."[25] In summary, one could say mules are an exception within

[23] Louis 1961: "Tous les autres produits de ces *mélanges* peuvent, on le sait, s'accoupler à leur tour entre eux" (88, emphasis mine).

[24] *GA* 2.7, 746b20–747a22. [25] *GA* 2.8, 747a23–25.

infertile animals because their infertility is not an individual exception but a characteristic of the species.

At first sight, Aristotle's obvious interest in the particular case of mules, to which he surprisingly devotes far more space than to hybridity in general, could be explained by this status of double exception, both within hybrids and within sterile animals. Indeed, Aristotle's interest in exceptions is characteristic of his scientific method, but in fact there is here a more general issue at stake than just mules, which is the relation between hybridity and sterility, and hence the limitations on hybridity: for Aristotle, mules are an exception and hybrids are usually fertile, so there is no causal relation between hybridity and sterility, and this latter does not work as a limitation on the former; but what calls for the long discussion about mules in *GA* 2.7–8 is a number of theories stating that mules' sterility is due to their being hybrids, and it is to three of them that Aristotle is replying here. Aristotle attributes the first two to Democritus (I) and Empedocles (II), while the third is a more general argument given without any precise attribution (III).

Let us begin with (III), described by Aristotle as an ἀπόδειξις λογική, which here can mean either a "logical" or "notional" demonstration, since it gives an explanation of mules' sterility not based on any physiological reasons but on the sole notion of what a mule is and is not.[26] Here are the premises: (1) from a male and female of the same species is born a male or female of the same species;[27] (2) male and female mules do not differ in species;[28] (3) a mule is a hybrid of horse and donkey;[29] (4) from parents of different species is born an offspring of different species than theirs.[30] These premises allow to explain mules' sterility by a *reductio ad absurdum*: if (hyp.) mules were not sterile, they would give birth to an offspring either [a] of another species or [b] of the same species; but [a] is impossible because according to (1) and (2) two mules should give birth to a mule and not an animal of another species; and [b] is impossible from (3) and (4) because a mule is a hybrid and thus cannot be of the same species as its parents;[31] so (not hyp.) mules have to be sterile.

Aristotle rejects this whole demonstration, but not because of the obvious logical flaw coming from too restrictive a reading of premise (3): "a mule is a hybrid of horse and donkey" is taken to mean that a mule can *only* be born of a horse and donkey, which begs the proposed conclusion that it cannot be born of two mules. One way to account for this restrictive

[26] Le Blond 1939: 210. [27] *GA* 2.8, 747b30–32. [28] *GA* 2.8, 748a1–2. [29] *GA* 2.8, 748a2.
[30] *GA* 2.8, 748a6–7. This premise is given at the end, after the conclusion. [31] *GA* 2.8, 748a3–7.

reading would be to consider that being a member of a species implies a given mode of generation. This is a quite Aristotelian idea since it is used at the beginning of *GA* in an argument (III′) about spontaneous generation. In order to prove that animals born of spontaneous generation cannot reproduce by sexual intercourse, Aristotle adduces the same alternative: the offspring should be either [a′] of another or [b′] the same species as its parents; but in case [a′], reproduction would imply a change in species which could be repeated indefinitely at each generation, and this is impossible because "nature avoids the infinite"; and in case [b′], the offspring, being of the same species as its parents, should have been born in the same way, i.e., by spontaneous, and not sexual, generation.[32] In reducing [b′], Aristotle seems to recognize that species and mode of generation cannot be separated, which could ground the restrictive reading of (3).

Although Aristotle would thus have a good reason to accept this reading of (3), he does *not* do so, not for the logical reason that it begs the question, but rather for the biological reason that it could be generalized to all hybrids, making all of them sterile by definition: if mules cannot reproduce because they can be born only of a horse and a donkey, then by the same line of argument all members of any hybrid species should be sterile because they could be born only of parents of a different species than their own. Reproduction would then be limited to animals of the same species as their parents', whereas animals of a different species than their parents' would not be able to reproduce. But to Aristotle this consequence is just contrary to the well-documented fact of most hybrids' fertility, mules being just an exception; so the whole "logical demonstration" of mules' sterility is "false, since many animals that are not born of the same species are fertile, as was previously said."[33]

Accordingly, Aristotle seems to accept for hybrids what he refused for spontaneous generation, i.e., some variation in generation mode inside a given species: the first generation of hybrids is born of parents of different species than theirs, whereas in later generations parents and offspring are of the same hybrid species. This would seem to contradict argument [III′] above where Aristotle rejects [b′] by denying the possibility of change from

[32] *GA* 1.1, 715b7–16.

[33] *GA* 2.8, 748a12–13: "οὐκ ἀληθὲς δέ, ὅτι πολλὰ τῶν μὴ ὁμοειδῶν γενομένων γίνεται γόνιμα, καθάπερ ἐλέχθη πρότερον." I do not think ἐξ should be added before ὁμοειδῶν as does Peck followed by Drossaart Lulofs: the issue here is not about "not being born of parents of the same species" but about "not being born of the same species as one's parents," which is what I think μὴ ὁμοειδῶν γενομένων means.

spontaneous to sexual generation within the same species. Yet one could argue in defense of Aristotle that for hybrids the generation mode remains in fact the same: only the parents' species vary, being different for the first generation and the same for following generations, so that all one has to suppose is a similar action of sperm on menses between hybrids of same species and between animals of different species at the first stage; which is clearly far easier to accept than a radical change from spontaneous to sexual generation.

Besides rejecting (III) by what he considers the evidence of fact (most hybrids are fertile), Aristotle also criticizes the method behind the argument. What Aristotle questions in this "logical demonstration" is not of course the use of logic in natural science, but rather the tendency to argue from principles that are too general and consequently unrelated to the issue at stake.[34] For example, if geometrical demonstrations are to be taken as a model of reasoning, it is not only by their strict logical consistency, but also because they are properly based on geometrical principles.[35] It should be the same in any science, and for biological problems all reasoning should be based on precise observation of each species' characteristics. What is wrong about (III) is that it takes no account at all of mules' or horses' or donkeys' observed peculiarities, but works on general principles (1) and (4), namely that animals of the same species give birth to another one of the same species, and conversely that animals of different species give birth to one of another species than theirs. For Aristotle, such exceedingly general arguments are void and say just nothing about their proposed object.[36]

Democritus' and Empedocles' Theories about Mules

Although the origin of (III) is not clearly stated, the analogy in structure and formulation with Aristotle's own argument (III′) about spontaneous generation could suggest that it is not necessarily an attested position outside the current discussion of *GA* 2.7–8: it could rather be an objection to his own theory constructed and immediately refuted by Aristotle for the sake of dialectics. This is not the case of the first two theories about mules' sterility rejected by Aristotle, since he attributes them respectively to Democritus (I) and Empedocles (II). Even if (I) and (II), unlike (III), are based on physiological arguments rather than general principles, for

[34] *GA* 2.8, 747b28–30. Cf. Cerami 2015: 38–40. [35] *GA* 2.8, 748a9–11.
[36] *GA* 2.8, 748a7–8 and 11.

Aristotle they have just the same failing as (III): they take no account of mules' specific characteristics but give a common explanation of sterility that applies, or could apply, to all hybrids.

Thus, Aristotle reports that according to Democritus "ducts are destroyed in mules' uterus because from the start they are not born from animals of the same genus."[37] A precise interpretation of this argument is difficult since there seems to be no other testimony about it, making us entirely dependent on Aristotle and consequently unable to distinguish between his report of Democritus and what Democritus may have really said or thought. So here (I) refers to the former, and as is often the case with Aristotle's reports of his predecessors, Democritus' theory can only be grasped through the Aristotelian filter of (I).

In spite of this uncertainty, one can suppose with some probability that the ducts mentioned here have the function of conveying the female seminal matter to the uterus, so the explanation would be quite similar to that previously given by Aristotle himself for woman's sterility, namely that "ducts from where the residue is secreted are troubled."[38] But the point is of course that according to (I) the alteration of ducts is not an individual variation but a common characteristic due to the parents being of different species. There is no precise explanation of this causal link between difference of parents' species and destruction of genital ducts, and indeed it would be quite difficult to imagine what such an explanation could be. Rather, (I) seems to depend on a more general conception that crossbreeding somehow alters the original species, an alteration that sometimes may go to the point of deformity or bring about some deficiency. Besides being based on evident observation, this conception of hybridity as a factor of alteration is invoked by Aristotle elsewhere, as we will see in the last part of this paper. But why does (I) precisely locate this alteration in genital ducts? Apart from sterility itself, Democritus may have tried in this way to explain observations about the lack of menses in mules. Aristotle mentions several times such observations,[39] but significantly he explains this absence of seminal matter without reference to hybridity: according to him, it is solely due to the big size of mules' body, the growth of which consumes all the residues that should form menses.[40]

If Aristotle avoids any explanation that would make hybridity the cause of the lack of menses or of sterility itself, it is because his only argument against (I) is first to generalize its theory to other hybrids, and then to

[37] *GA* 2.8, 747a29–31 [= DK 68 A 151]. [38] *GA* 2.7, 747a7–13.
[39] *GA* 2.8, 748b24; *HA*, 6.18, 573a15–16. [40] *GA* 2.8, 748b19–20.

contradict it by what he considers to be a fact, i.e., the fertility of most hybrids.[41] One cannot know for sure if Democritus would agree with the first step of generalization: did he really give this explanation also for other hybrids, or only for mules, the generalization being then a polemical device of Aristotle? In any case, Aristotle's stance is to hold that if hybridity were the cause of such a deformity as to cause sterility, this should happen in other species than mules, which according to him is just contrary to fact. The induction from mules to other hybrid species is quite weak and maybe Democritus would have objected to it, but what makes (I) liable to such a generalization is once again its lack of consideration of mules' proper characteristics: there is nothing in (I) that could limit its explanation of sterility to the case of mules and this is what Aristotle takes advantage of, possibly by simplifying Democritus' position in his own formulation of (I).

Undeniably, despite its being refuted as too general, (I) has a strong Aristotelian flavor since it seems to rely on principles and observations familiar to Aristotle. Things are quite different with Empedocles, whom Aristotle explicitly blames for lacking clarity[42] as if he had failed to make any Aristotelian sense out of his theory. As a matter of fact, one cannot deny that Aristotle's testimony about Empedocles, here referred as (II), is rather obscure, even if all the blame should not be put on Empedocles' original theory: whatever this may have been, it is obscured in (II) by the lack of clear distinction between Aristotle's report of the theory and his refutation of it, and incidentally also by many textual problems originating in a corrupt manuscript tradition. Nonetheless, the core of Empedocles' theory according to (II) is fairly clear: when semen from both the male and female is soft, the mixture produced becomes something compact by the mechanical combination of hollow and solid parts in seminal matter coming from both parents;[43] now semen coming from both horse and donkey is soft, so the product of their union is so compact that it becomes sterile.[44] What seems a little strange about this explanation is that it is based not on physiological characteristics of adult mules, but rather on the physical texture of the semen mixture from which they are born; of course, the relation between sterility of adult mules and consistency of the mixture between horse and donkey semen remains completely unaccounted for, and here Aristotle may be right in blaming Empedocles' lack of clarity.

[41] *GA* 2.8, 747a31–34. [42] *GA* 2.8, 747a26–27. [43] *GA* 2.8, 747a34–b3 [= DK 31 B 92].

[44] *GA* 2.8, 747b13–14 and 17–19. In the latter passage, I choose Platt's correction of ἕν in πυκνόν before τι γίνεται, instead of adding πυκνόν after γίνεται as does Peck followed by Louis and Drossaart Lulofs.

Yet Aristotle's refutation, without targeting this obviously weak point, concentrates of Empedocles' own reasoning about semen consistency. Aristotle's objection is that if a mixture between soft semen from both parents produced a sterile offspring, then also horses should be sterile[45] because semen from both stallion and mare is soft.[46] To demonstrate this point, Aristotle argues from the fact that crossbreeding can happen between stallion and she-ass as well as mare and he-ass.[47] This argument is not completely clear because of Aristotle's very concise formulation and philological problems in establishing the text, but the line of reasoning seems to be as follows: let us call μ (as Greek μαλακόν) the property *softness of semen* and let us take it as belonging to the stallion; if only crossbreeding of stallion and she-ass were possible, one could say property μ in the stallion is cause of sterility in their offspring (S) because it is identical to μ in the she-ass;[48] but in fact crossbreeding is also possible the other way round, namely between mare and he-ass, and the offspring born of it is also sterile, which means that it comes from a mixture of soft semen from both parents, and accordingly that μ also belongs to the mare and he-ass; so μ in the stallion meets another μ not only in the case of crossbreeding, but also when mating with a female of same species,[49] and consequently their offspring should be sterile.

[horse]		[donkey]		
♂ (μ)	+	♀ (μ)	→	S
+		+		
♀ (μ)	+	♂ (μ)	→	S
↓		↓		
S		S		

If this interpretation of (II) is correct, Aristotle's point is to show the contradiction between Empedocles' explanation and the reciprocity of hybridity between both sexes of each parent species. Yet another reason of Aristotle's criticism against Empedocles is that he is supposed, like Democritus, to have given a common explanation of sterility for all hybrids.[50] Once

[45] *GA* 2.8, 747b19–20. [46] *GA* 2.8, 747b15–16. [47] *GA* 2.8, 747b11–12 and 16–17.

[48] *GA* 2.8, 747b20–22. I keep the reading ὅμοιον of manuscripts because the reason for sterility is that the semen from both species is "identical," i.e., soft; I think this crucial point is missed with Platt's correction ἀνόμοιον followed by Peck and Louis, or with οὐχ ὅμοιον chosen by Drossaart Lulofs from an isolated Greek manuscript and the Arabo-Latin tradition against more reliable Greek manuscripts (notably ZPS).

[49] *GA* 2.8, 747b22–23. [50] *GA* 2.8, 747a28–29.

again, as we depend entirely on Aristotle's testimony, we cannot know for sure if Empedocles used the same argument for other species, and nothing in (II) actually indicates such a thing. Rather, just like in the refutation of (I), it may be again that Aristotle himself generalizes to other hybrids an explanation based on a causal link between mules' sterility and their being hybrids. In Aristotle's eyes, Democritus and Empedocles are similar in that they both relate sterility and hybridity, although they actually do it in opposite ways: for Democritus, it is difference which causes deformity and deficiency, whereas for Empedocles it is sameness, namely the common property of softness in semen coming from both species, which makes the mixture too compact and sterile.[51]

Hybridity and Heredity

Unlike theories (I)–(III), Aristotle's explanation of mules' sterility cannot be generalized to other species and remains strictly specific to this case: as Aristotle puts it, "it is by considering what belongs to horse's or donkey's genus that one may best grasp its cause."[52] One might find a little strange that sterility in *mules* should be explained by particular features of *horses* and *donkeys*, but it probably means that for Aristotle accounting for a hybrid's characteristics implies going back up the causal chain and searching their origin in the crossed species: just as children look like their parents, so hybrids' peculiarities must be somehow similar to those of the original species. Here Aristotle establishes a relation between mules' sterility and low fertility of horses and donkeys: the fact that these two species are uniparous already shows this lack of fecundity, which Aristotle explains by different factors such as the small quantity of menses in mares or coldness of semen and menses in respectively he- and she-ass.[53] In this way, horses and donkeys "have a natural tendency to sterility,"[54] and when they mate this predisposition is reinforced and turns into actual sterility in their offspring. But why is it reinforced? Even if Aristotle is not very clear on this point, he seems to resort to two distinct explanations, one for each of the crossed species: (1) horses are almost, but *not* sterile because seminal

[51] There may be a relationship between this infertility caused by sameness and the crucial role of difference in Empedocles' theory of reproduction: Aristotle reports that Empedocles used the metaphor of the "symbol" (a piece of pottery broken in two parts that fit perfectly), to express the fecund complementarity between both sexes, each of them providing a different part of the embryo (*GA* 1.18, 722b10–13); of course, the two parts of the symbol cannot fit if they are identical (*GA* 1.18, 722b15–16).

[52] *GA* 2.8, 748a14–16. [53] *GA* 2.8, 748a17, 748a20–21 and 748b3. [54] *GA* 2.8, 748b8.

secretions are just warm enough to allow generation, but this is no longer the case after the mixture with donkeys' colder secretions, which makes mules sterile[55]; (2) donkeys are almost sterile and consequently give birth to almost sterile offspring even in natural generation between them, and this very low fecundity turns into sterility in mules because they are from an unnatural union.[56]

Although Aristotle provides no further details about these two explanations, they give us some hints about hybrids' specific heredity. Argument (1) can easily be brought back to the general principle of similarity between parents and offspring: hybrids being born of two different species, they naturally have some intermediary characteristics between them; for instance, mules' seminal secretions will be colder than horses' because this feature of coldness also belongs to the other species from which they originate. Argument (2) is more of a problem, for in his discussion of (I), (II), and (III) Aristotle seems very cautious to avoid postulating any direct causal link between the fact of hybridity itself and sterility: his main strategy against these three theories consists in generalizing this supposed link, whatever it may be, and showing that it is proved false by most hybrids' fertility; so how can he say now that the mere fact of being born from an unnatural union lowers fecundity and turns it into sterility?

What could explain this weakened fecundity is a general tendency to the weakening of crossed species' features. One example may be found in the case of Indian dogs, born from a female dog and a wild beast called by Aristotle a τίγρις:[57] a passage from *HA* states that such hybrids of the first generation are still wild beasts and hybridization has to be repeated to obtain the race of "Indian dogs,"[58] as though the tiger's wildness were progressively diluted by the domestic nature of the mother dog. Such a dilution of the original species' feature could also help us to understand a rather surprising statement made in *GA* according to which hybrids tend to become more feminine one generation after another: "When a male and a female not belonging to the same genus mate together [. . .], first is born an animal common to those two as far as resemblance is concerned [. . .], but over time, a generation begetting another one, eventually they end up feminine in form (τὴν μορφήν)."[59] This process of feminization is not entirely clear because there is an ambiguity about the feminine "form"

[55] *GA* II 8, 748b11–14. [56] *GA* 2.8, 748b14–19.

[57] See the section "Conditions of Hybridity" and n. 11.

[58] *HA* 7.28, 607a3–6: "φασὶ δὲ καὶ ἐκ τοῦ τίγριος καὶ κυνὸς γίνεσθαι τοὺς Ἰνδικούς, οὐκ εὐθὺς δὲ ἀλλ' ἐπὶ τῆς τρίτης μίξεως· τὸ γὰρ πρῶτον γεννηθὲν θηριῶδες γίνεσθαί φασιν."

[59] *GA* 2.4, 738b27–34.

(μορφή) toward which it tends. Is it the morphology of the female hybrid, which would mean that over generations hybrids, including males, incline to have more and more feminine traits? Or is it rather the morphology of the mother from which the first generation was born, which would mean that little by little both male and female hybrids tend to get closer to the species of the first generation's mother rather than that of the father?

Without ruling out completely the first interpretation, the second seems more likely for several reasons. First, there is a clear opposition in Aristotle's text between this tendency to feminine morphology in later generations and the common resemblance to both parents' species at the first step, which strongly suggests that the process of feminization has to be referred back to the species of the female from which first generation hybrids were born: at first there is a balance between the father's and the mother's species, but gradually this balance is lost and the latter prevails on the former. The text about Indian dogs in *HA* also suggests that the domestic nature of the dog mother prevails over the wildness of the "tiger" father, not at the first generation, but at the "third mixture;" yet this case is not completely decisive since the gradual domestication could be attributed either to the aforesaid process of feminization or to repeated hybridization with a dog. A more conclusive example is incidentally given in another text of *HA* where Aristotle mentions that crossbreeding between horse and donkey gives an offspring more similar to the mother, as if there was, right from the first generation, a feminization: "Size, form, and strength of the offspring resemble more the female."[60]

In Aristotle's general theory of generation, form/species (εἶδος) comes from the male and feminization of morphology (μορφή) is equivalent to a loss of form, due to some resistance to it from the matter provided by the female. If hybrids' heredity is characterized by progressive feminization, it is probably because difference of species at the first step creates some kind of resistance to transmission of form: in the first generation, the loss is still minor so the hybrid resembles both parents, but over generations there is a cumulative effect which weakens specific features coming from the original male and makes them closer to the original female. Increased resistance from matter and loss of form are also what may account, in the case of mules, for their weakened fecundity compared to their parent species. So, for Aristotle, hybridity is not in itself a direct or sufficient factor of sterility, but it adds some material resistance to the transmission of form:[61] of

[60] *HA* 6.23, 577b10–11.

[61] This resistance is also suggested by the frequency of miscarriages (*HA* 6.23, 577b5–7).

course, the ability to reproduce is an essential part of form and its decrease, combined with already low fecundity in horses and donkeys, makes mules sterile. In fact, sterility is just a matter of degree, and there is a kind of continuum between low but sufficient fecundity in horses or donkeys and low but insufficient fecundity in mules. When Aristotle says that mules are even more sterile,[62] this means by conversion that they are less fertile, so here fecundity and sterility are relative, not absolute: in a way, mules *are* fertile, and indeed Aristotle mentions cases of pregnant females and begetting males; but of course this fertility is very low, since these cases are extremely rare and pregnancy does not go to term.[63]

Conclusion

Aristotle manages to explain why mules are sterile in a delicate poise between particular and general: unlike Democritus' and Empedocles' theories or the "logical demonstration" of mules' sterility, Aristotle's explanation cannot be applied to other hybrids since it is based on a peculiarity of their parent species, namely the low fecundity of horses and donkeys; yet, this explanation also has a general bearing since the weakening of this low fecundity into sterility allows us to understand the specific features of hybrids' heredity in the broader framework of Aristotle's theory of generation.

The question that remains is why Aristotle considers this case so important as to devote it most of his discussion about hybridity; this is all the more surprising that he could have used mules' example to consider sterility as a limit of hybridity, but actually takes the opposite view. From a historical point of view, the main reason why Aristotle goes at length to explain mules' sterility is the obvious existence of some controversy about it, possibly originating in agrarian knowledge but which had already become a philosophical matter, as is clearly shown by the references to Empedocles and Democritus. And from the theoretical point of view, if Aristotle devotes so much space to the case of mules without making sterility a restrictive condition of hybridity, it may be for some or all of the following reasons: first, most likely Aristotle did not think that hybridity posed any threat to the general consistency of his fixist biology, because it is sufficiently limited by a set of rather strict conditions on the parents' species such as similarity in size and nature or equal gestation time; second, Aristotle's strategy of refuting general (or possibly general) explanations of

[62] *GA* 2.8, 748b17–18. [63] *GA* 2.8, 748b29–34; cf. *HA* 6.24, 577b19–23.

mules' sterility as contrary to fact shows that he felt compelled to do so by some corpus of observations attesting ability to reproduce in most hybrids; third, Aristotle's interest in the case of mules might have arisen precisely from its being an exception to the rule, i.e., something that has to be explained in order to suppress the threat it posed to the general assumption of hybrids' fertility.

For Aristotle, the scientific validity of a set of explanations depends on its ability to account both for what happens most of the time and for exceptions. The discussion of hybridity in *GA* 2.7–8 is quite a remarkable example of the way this Aristotelian science of exception works, always reconsidering and refining its own discoveries first from natural generation to the exceptional case of hybrids, then inside them to mules, which are an exception to hybrids' fertility, then inside mules to the few cases of fertility (an exception inside an exception inside an exception).

CHAPTER 9

Females in Aristotle's Embryology

Jessica Gelber

Introduction

How does Aristotle view the production of females? According to what I will call the "standard reading," Aristotle thinks that a female birth is a failure. There are several variations of this reading,[1] but the central claim that the various versions have in common is that a female is the result of a defective teleological process. For instance, Marguerite Deslauriers reports that Aristotle considers females to be "defective in the sense that if the form from the male parent had been transmitted as it should have been, male offspring would have resulted."[2] Karen Nielsen claims that Aristotle "conceives of individual female animals as the result of an incomplete, and therefore failed, teleological process."[3] According to Gareth Matthews, "males are supposed to be successes, and females failures."[4]

That Aristotle thinks that females are the results of thwarted causal processes is the prevailing view. Despite its prevalence, however, it is not a view Aristotle ever explicitly expresses in his biological account of animal generation. As we will see, that view blatantly contradicts what he does explicitly say about female births.

An alternative to the standard reading, offered by Devin Henry, paints a very different picture. On Henry's interpretation, the production of a female is not due to a failed causal process. Rather, the embryo's sex is

I would like to thank the editors for their suggestions for improvement, as well as the participants at the conference in Paris. I am also grateful to Kathleen Cook, Marguerite Deslauriers, Vanessa de Harven, Joseph Karbowski, Jim Lennox, Nathaneal Stein, Stasinos Stavrianeas, and Jennifer Whiting for their helpful written comments and discussion.

[1] Not all proponents of the standard reading, for example, think that females do not share the species-specific forms of their male counterparts. (Human women are, after all, still *human.*) Many scholars interpret Aristotle to be claiming that females realize that form in an impoverished or defective way. For a critique of the standard view in connection with Aristotle's ethics and politics, see Karbowski 2012.

[2] Deslauriers 1998: 159. [3] Nielsen 2008: 374.

[4] Matthews 1986: 23. See also Witt 2012: 102–4.

determined solely by "non-teleological necessity," which Henry glosses as the necessity "attached to the interactions between material forces which do not occur for the sake of their end (e.g., when cold air acts on the surface of a pond forming ice)."[5] According to this "alternative account," as I will call it, Aristotle's biological explanation is neutral with respect to the sex of the embryo; successful generation requires only that the offspring be one sex or the other.

This alternative account, as I will argue, also misrepresents Aristotle's view. Aristotle does not say that sexual determination is due merely to the interaction of material forces. The explanation Aristotle gives is more sophisticated – and less egalitarian – than "material necessity" suggests.

Here I will argue for a third option. Although I do not think the cause of sexual determination is "material necessity," I agree with Henry that, considered at the level of species-form transmission, the process of animal generation does not aim at producing males.[6] However, there is more than one level at which Aristotle considers generation. As I understand the details of the mechanism by which sex is determined – the motions and potentials (κινήσεις and δυνάμεις) that Aristotle discusses in *GA* IV – males are in a certain sense the "default" result.[7] But, as I will argue, the way in which Aristotle privileges the mechanisms for producing males is consistent with females not being teleological failures. This interpretation better coheres with the texts than either the standard reading or Henry's alternative account, and it avoids attributing conflicting views to Aristotle.

The Standard Reading

Because it is so common, I am going to assume that the standard reading is familiar. In broad brushstrokes, this reading ascribes to Aristotle the view that generation is "teleologically directed toward the production of male

[5] Henry 2007: 257, n. 18.

[6] I will assume here that the concept of form employed in Aristotle's embryology is the same, species-level form as he employs in his *Metaphysics*, at least as traditionally understood. If one thinks, like Balme 1987a, that the concept of form in biology is that of an individual (subspecific) form, which includes not only sex but also all of the father's inheritable traits, then not resembling the father in any respect (and not only being female) is a failure of form to be realized. For arguments against that "subspecific" interpretation of Aristotle's embryology, see Gelber 2010.

[7] I want to stress that by "default" I do not mean to suggest that Aristotle thinks that male replicas of the sire *should* or *are supposed to* result. Given that replicas almost never get produced, I do not think Aristotle could think that is what is supposed to occur. By "default" I mean merely to describe what would always happen, *per impossibile*, if we abstracted from actual conditions in which the causal process takes place. Ideal Gas Laws, for instance, would be descriptions of a "default" process in the sense I have in mind.

rather than female animals."[8] In order to assess the textual evidence for this reading, it is important to distinguish the claim that Aristotle views the production of females to be failures from the undeniable fact that he considers males – especially human men – to be better than their female counterparts. For, although Aristotle quite certainly thinks that females are inferior to males in many respects, it does not follow from this that the production of a female is a defective generative process. Evaluation of the causal process and evaluation of some product can, and often do, take place independently. For example, consider the process of learning a language, e.g., French. The goal of that process, we might assume, is the acquisition of French form, i.e., the ability to speak and read and understand French. This goal will be achieved so long as the student can speak French with an accent that falls within some range of acceptable ones. It might very well be the case that a particular accent, e.g., a Parisian one, is more highly prized than others, perhaps for reasons having to do with social status. Yet that superior accent need not be the goal of the teaching process, such that acquiring any other accent would constitute a failure. Similarly, male offspring can be better than female ones, though not be what the reproductive process aims to bring about. Consequently, the fact that Aristotle thinks males are better than females is not evidence that he also thinks females are the results of a failed causal process.

What about Aristotle's statements that human women are "like infertile (ἄγονον) males" (728a17), that a female is "like a disabled (πεπηρωμένον) male" (737a28), that female nature needs to be considered "like a natural mutilation (ἀναπηρία φυσική)" (775a15–16), and that the nature of human women is "nearly resembling (παραπλησία) that of children" (784a5)? Does this show that Aristotle thinks females are defective results?

Not necessarily. For, in Aristotle's biology, male and female are primarily differentiated by the ability each has to concoct residues and make their respective reproductive contributions.[9] And it is that ability to concoct that is at issue in each of those passages cited above. Males have a higher degree of heat, and so can fully concoct spermatic residues into semen, whereas females have a lower degree of heat, and so can only produce menses. In that respect, females are like mutilated (i.e., castrated) males and like boys, neither of which can fully concoct spermatic residues into semen. If we do not presuppose that generation aims at producing males, it is possible to read these passages more neutrally, in a way compatible with the view that successful generation merely requires that the offspring have

[8] Nielsen 2008: 374. [9] *GA* 4.1, 765b8–14; *GA* 4.1, 765b35–766a4; *GA* 4.1, 766a30–b4.

heat within some range such as to be able to make a reproductive contribution.[10] No doubt the male is superior along that dimension, and he is therefore "more perfect" or "complete."[11] Yet that superiority, and the female's corresponding "disability" with respect to concoction, need not imply that females are failures.

Another piece of putative evidence that Aristotle thinks males are not merely superior specimens but also superior *qua* effects of generation seems to be based on a misreading of certain passages. For instance, Aristotle does say (*GA* 4.3, 768a21–2) that it is "most natural" *that* a son also resembles his father. But Aristotle is not saying in that passage that is most natural for the offspring *to be* a son that resembles his father. For, as Devin Henry has correctly noted,[12] Aristotle says that it is most natural for the causal factors involved in determining male sex and those involved in conveying the father's traits *to master* and *be mastered* at the same time.

> It is, then, most natural that the [motion or potential or principle][13] *qua* male and that *qua* father master and are mastered at the same time [. . .]. For this reason, males for the most part resemble the father and females resemble the mother. (GA 4.3, 768a21–22, 24–25)

This passage claims that male sex and paternal resemblance usually go together, as do female sex and maternal resemblance. This is why sons often resemble their fathers and daughters resemble their mothers. This passage does not claim that it is most natural that the offspring be a male who looks like the father.

In short, there is a paucity of direct evidence in support of the idea that females are simply defective results. Moreover, proponents of the standard reading generally acknowledge that sex is not part of the form (εἶδος) of any animal kind.[14] If it were, males and females would differ in εἶδος, which Aristotle denies.[15] Assuming that generation aims, strictly speaking, at producing an organism the same in εἶδος as the parents, then producing a male is not the goal of generation. Now, it might not follow from this

[10] This way of construing sex would make sense of Aristotle's use of the comparative "more womanish" (γυναικικώτερα, 766b32), which is otherwise a bit odd. (The text is ambiguous between attributing that adjective to the parents and the female offspring.) The "more womanish" individuals would be the ones that have less heat. The hotter the organism, the more male it is, and colder the organism, the less male and more female it is.

[11] See *PA* 2.2 on the correlations between heat and "perfection" across different kinds of animals.

[12] Henry 2007: 258.

[13] A feminine noun needs to be supplied here. It is clear from context that it refers to the causal factors involved in sexual determination and inherited traits.

[14] See, e.g., Deslauriers 1998 and Matthews 1986.

[15] *Meta.* 9.9, 1058a29ff. Cf. *GA* 1.23, 730b33–731a1.

that females are not defects, since a female could have the same form or essence as her male counterparts but still be defective. However, the respect in which a female is defective needs to be made precise. For instance, females are defective with respect to the ability to concoct spermatic residues. But being defective in that respect does not amount to being defective results of the generative process.

Nowhere in the *GA* does Aristotle explicitly say that the production of females is a failure in the sense maintained by the standard reading. That standard reading, consequently, lacks clear textual support.

In addition to lacking textual support, this reading is in tension with what Aristotle says about the production of females. For, Aristotle thinks there is a purpose or goal achieved by the production of females. A sustained discussion of the purpose for which there are females is found at the beginning of *GA* 2.1. The passage opens with a statement of the phenomenon to be explained:

> The female and the male, that they are principles (ἀρχαί) of generation, has been stated earlier, as well as what is each's potential (δύναμις) and definition (λόγος τῆς οὐσίας). Why one comes to be and is female and the other is male, as from necessity and the first mover and what sort of matter, the account proceeding must try to say. But as on account of the better and the cause for the sake of something, the account has a higher principle. (*GA* 2.1, 731b18–24)[16]

Here, as elsewhere in Aristotle's natural science, he distinguishes the efficient and the material causes – the causes "as first mover and what sort of matter" – from the final causes – "the better and cause for the sake of something." So, it is natural to read this as introducing a teleological explanation for the same phenomenon that will be given a material and efficient causal account later. In fact, when we get to *GA* 4, where he explains how sex is determined during embryonic development, he describes what he is explaining with the same language: At the end of *GA* 4.1, Aristotle says that he has explained "on account of what cause the one comes to be female and the other male" (766b27–28).[17] What exactly

[16] I am going to bracket questions both about translation and interpretation of the last clause of this quote, in particular questions about how ἄνωθεν should be rendered. For various suggestions see Peck 1942 (who takes it as a reference to the prime mover), Balme 1992 (who takes it more literally, as a reference to what was said previously), and Coughlin (who argues that this is a reference to a principle "higher" in that it is derived from Aristotle's general account of change, rather than biology).

[17] The only difference between the two phrases is the insertion of "cause" (αἰτία) in 4.1, so that it reads "on account of what cause" instead of "on account of what."

is the phenomenon for which he gives a teleological explanation in *GA* 2, and then a material-efficient explanation in *GA* 4?

The stated *explanandum* in both passages is why "the one comes to be female and the other male," a fact for which in book 2 Aristotle provides a series of interrelated, teleological explanations. First, he explains why there is always a genus (γένος) of humans and animals and plants:

> For, since some things are eternal and divine, while others are able both to be and not be, and the noble and divine is in accordance with its nature always a cause of the better in the things able [to be better], and the non-eternal is able also to be and not be and partake in the worse and the better, and soul is better than body, ensouled is better than soul-less on account of the soul, to be is better than not be, and to live is better than to not live, on account of these causes there is generation of animals. For, since the nature of such a kind is unable to be eternal, in accordance with the way it is able, in that way what comes to be is eternal. Well, then, in number it is not able. For, the being of these is in the particular. If it really were such [i.e., able to be eternal in number], it would be eternal. But in form (εἴδει) it is able [to be eternal]. Wherefore, there is always a kind of humans and animals and plants. (*GA* 2.1, 731b24–732a1)

Although the Greek is compressed, the overall meaning is fairly clear. As a general metaphysical principle, being is better than not being. For living things, which exist in virtue of being alive and so ensouled, this entails that living is better than not living. But such living substances are perishable, and as such, they obviously cannot be eternal "in number." Instead, they achieve their eternal existence through the persistence of their kinds: this is how perishable substances can achieve the only type of eternal existence available to them – to be eternal "in form." This type of eternal existence is that for the sake of which the kinds persist.[18]

Any kind of living organism's persistence, moreover, is dependent on the continued generation of individual members of the same kind. Generation, in Aristotle's view, typically requires both male and female principles.[19] So, both of those principles must exist. That there are male and female principles, even mixed as they are in plants, is a second *explanandum.*

> Since the principle of these is the female and the male, for the sake of generation the female and the male should be in those having [male and female principles]. (*GA* 2.1, 732a1–3)

[18] For a slightly different construal of the first of these three "hierarchically related" teleological explanations, see Lennox 2001a: 133ff.

[19] Spontaneously generated organisms, for example, might be an exception. It appears that these have causal principles corresponding to (κατά) the male and female ones. See *GA* 3.11.

Third, he explains why it is that the male principle is separate from the female one in most animal kinds: It is better that the "more divine" and superior male principle be separate from the inferior female principle.[20]

> Since the primary moving cause, to which the account and form (εἶδος) belong, is better and more divine in nature than the matter, better also is the separation of the superior from the inferior. On account of this, among those able and to the extent that they are able, the male is separate from the female. For, the principle of movement, in virtue of which "male" is applied to generated things, is better and more divine; matter [is that in virtue of which] "female" [is applied]. (*GA* 2.1, 732a3–9)

Taken together, these three explanations amount to an explanation for why some animals are male and some are female, which Aristotle makes explicit when summarizing the discussion later on:

> And concerning male and female, on account of what cause each exists, has been said. (*GA* 2.1, 732a24–25)

As I read this passage, Aristotle is giving a teleological explanation, one citing the end for the sake of which some animals are produced male and some female. So it seems as though he thinks that there being females as well as males is for the sake of something. Moreover, that thought is echoed in his remarks in *GA* 4.3:

> For, not resembling the progenitors is already a monstrosity in a way.[21] For, nature in these cases has departed from the type[22] in a certain way. The primary source[23] [of the departure from type] is the female coming to be and not the male. But, on the one hand, this [primary source of

[20] I am going to put aside questions about why it is better for the superior principle to be separate from the inferior one.

[21] The qualification "in a way" (τρόπον τινά) here and in the next sentence is important: Aristotle does not think that all nonreplicas are, strictly speaking, monstrosities. He goes on to account for monstrosities in the following chapters of Book 4, after completing the account of sexual differentiation and familial resemblance.

[22] As David Lefebvre (2014: 216–7, note 26) has noted, γένος (here translated "type") is unlikely a reference to the γένος of human beings, and is more plausibly a reference to the child's parental (and, more specifically, paternal) lineage or clan.

[23] The ἀρχὴ δὲ πρώτη at the beginning of this sentence tends to be read as implying that the birth of a female is the first in a series of successive deviations from the ideal – i.e., male resemblance – that culminates in monstrosities. This is not the sense, as far as I can tell, of the corresponding sentence in Michael Scot's Latin translation of the Arabic: "*Non est res nisi generatio feminae, non generatio maris, et hoc indiget natura necessario, quoniam debet salvari genus animalis in quo sunt mas et femina distincta.*" Though I will not pursue this here, I think it is possible to read this as the claim that failure to resemble the sire would not occur if not for the fact that sometimes females come to be. That is, females would be the "primary source" of departure from type in the sense that the female's involvement is the primary cause: if there were not females involved, and the male were generating by himself, exact resemblance would occur.

> departure from type] is necessary by nature. For, the kind of those separated into female and male must be preserved. Since it is possible for the male not to master either because of youth or old age or some other such cause, the female offspring among those animals necessarily comes to be.[24] A monstrosity, on the other hand, is not necessary for the sake of something and cause as end, but accidentally necessary, since one must grasp the principle from there (ἐντεῦθεν). (*GA* 4.3, 767b5–15)

Failure to resemble one's progenitors is, in a way, a monstrosity, and the "primary source" of such failure to resemble one's progenitors is the birth of a female. But despite the birth of a female being the primary source of departure from type, the birth of a female is necessary "by nature." While Aristotle does not say explicitly here that to be necessary "by nature" is to be for the sake of something, the μέν . . . δέ construction of the passage suggests that this is what he means. For, he says that monstrosities are only necessary "accidentally" and do not come to be for the sake of anything. Females, on the other hand, are naturally necessary since "the kind separated into males and females must be preserved" (*GA* 4.3, 767b9–10). Given the contrast drawn in this passage between females and monstrosities – namely, that one is and the other is not for the sake of something – it appears that Aristotle thinks that females are produced for the sake of something.[25]

So, Aristotle thinks that females are "by nature" necessary and – unlike monstrosities – are for the sake of something. If the standard reading is correct, Aristotle also views females as failures. That reading renders females the accidental results of a causal process aiming to produce males. However, these two views are inconsistent: one and the same thing cannot be both for the sake of something and also a failure.[26]

One and the same thing can, of course, be both *beneficial* and also a failure. So, if all Aristotle means by saying females are for the sake of something is that females are beneficial or purpose-serving, there would be nothing incoherent about them also being accidental results or even failures. Surely there are plenty of things that get used for some purpose or turn out to be beneficial but that nevertheless are the results of a failed causal process aiming at some other end.

[24] Assuming Aristotle does not subscribe to a principle of plentitude, the inference here, "since X is possible, Y is necessary," is strange. I return to this below in the section entitled "A Third Option."

[25] Nielsen 2008: 376–7 claims that in this passage Aristotle likens females to monsters.

[26] On this point, I completely agree with Devin Henry: "The standard reading thus saddles Aristotle with the paradoxical view that females are present in the species for the sake of something even though no particular female comes to be present for the sake of anything but is merely an accidental result of a process aimed at generating males" (Henry 2007: 261).

This deflationary sense of being for the sake of something – i.e., being merely beneficial – cannot be what Aristotle has in mind, however. First, Aristotle takes it to be a substantive and controversial thesis that natural phenomena occur for the sake of something. But the idea that natural phenomena can turn out to be beneficial or serve some purpose is completely uncontroversial. Second, results of luck such as a chance meeting between debtor and lender or finding a treasure buried in your backyard would, in this deflationary sense, count as being for the sake of something. Such accidental yet beneficial results, however, are explicitly contrasted with those natural results that are genuinely for the sake of something.[27] This deflationary way of being "for the sake of something" that involves merely being beneficial or useful is, in my view, too thin.

The Alternative Reading

The textual evidence in support of the standard reading is not, as I have argued, very compelling. At any rate, Aristotle does not say explicitly anywhere that the goal of generation is a male that looks like its father. Furthermore, on the standard reading, Aristotle holds grossly inconsistent views about the production of female offspring. If Aristotle has a consistent position, he cannot think that females are failures.

An alternative to the standard reading has recently been offered by Devin Henry. In Henry's view, generation in no way aims to produce males. Rather, on his interpretation, the sex of the embryo is "determined entirely by nonteleological forces operating through material necessity."[28] Henry's interpretation, consequently, avoids the inconsistency that the standard reading faces. However, Henry's alternative account does not accurately describe Aristotle's view of the production of females for the following two reasons. First, the cause of sex determination is not merely material necessity. Second, there is an asymmetry between Aristotle's account of male and female births that the alternative account fails to appreciate.

In order to see why the description of the causal process as "material necessity" is an oversimplification of Aristotle's explanation of sex determination, we need to turn our attention to the details of *GA* 4.1–3. Henry

[27] For example, Aristotle's strategy in *Phys* 2.8 for arguing against an opponent of natural teleology is to show that "purpose serving" natural phenomena cannot be due to chance and, consequently, are finally caused.

[28] Henry 2007: 268.

is right that in Aristotle's biological explanation the determination of females involves a "deficiency of spermatic heat."[29] For, lacking enough heat renders the principle from the male incapable of conveying his sex to the offspring.[30] This is not, however, the full explanation. What Aristotle says is that if "the patient is not mastered on account of a deficiency of the concocting and moving power or on account of the bulk and coldness of the thing being concocted and determined" (*GA* 4.3, 768b25–27), there is a "change into the opposite":

> Whenever the principle should not master (μὴ κρατῇ) and neither is able to concoct on account of a lack of heat, nor able to lead to its own form but in this respect is defeated, it is necessary to change into the opposite. And the opposite of male is female, and in this respect one is male and one is female. (*GA* 4.1, 766a18–22)

In *GA* 4.1, there is not much more said about what happens when the "change to the opposite" occurs. The account of sexual determination in *GA* 4.1 is given almost exclusively in terms of defeat: there is a "lack of heat," the male principle is "unable to concoct" and "gives way." However, at the beginning of *GA* 4.3 Aristotle says that not only the sex of the offspring but also its inherited characteristics are due to the "same causes," and then repeats once again at the end of the discussion (at 769a1–6) that he has discussed these phenomena together. And within the discussion of the causes of offspring resembling their parents and ancestors in *GA* 4.3, Aristotle uses the same terminology to describe the causal mechanism. There in *GA* 4.3 we again find the language of "mastery" and "failure to master," and "change into the opposite." For example, an offspring bearing resemblances to the mother or maternal ancestors is due to a failure to gain mastery, and a subsequent change into the opposite.

In that *GA* 4.3 account, Aristotle gives far more details about the process than he does in *GA* 4.1. In particular, in *GA* 4.3 Aristotle appeals to "motions and potentials" (κινήσεις καὶ δυνάμεις) in the spermatic fluids by

[29] Henry 2007:262.

[30] The role heat plays in determining the embryo's sex accounts for the purported preponderance of female offspring from parents younger or older than their prime (*GA* 4.2, 766b28–31): in the young, the heat is not yet perfected, and in the old it is leaving. This is also why male offspring are more common when the parents copulate during northerly than southerly winds (*GA* 4.2, 766b34–767a1): when the wind is in the south, bodies are more fluid and so abundant in residue, and an abundance of residue is more difficult to fully concoct (*GA* 4.2, 766b35–36). And shepherds say that the direction the copulating animals are facing also affects the sex of the offspring (*GA* 4.2, 767a8–11). This is to be expected, Aristotle thinks, since such small differences cause differences in heat and coldness (*GA* 4.2, 767a11–12).

which particular traits are passed along to the offspring.[31] There are, Aristotle says, motions and potentials corresponding to the father at varying levels of generality. There are motions and potentials not only *qua* member of a certain species (such as human), but also *qua* male, *qua* individual male, and *qua* animal.

> I speak of each potential in this manner. The generator is not only a male but also such a male, e.g. Coriscus or Socrates, and he is not only Coriscus but also human. And in this sense, some things that belong to the generator are closer and some further *qua* generator and not accidentally, such as being literate or someone's neighbor [. . .]. For this reason, motions are present from the potentials in the spermatic fluids of all such things. (*GA* 4.3, 767b23–29, and 35–36)

According to Aristotle's account, when the offspring resembles its father in a certain respect, for example with respect to the shape of the nose, that offspring's nose shape is the *per se* result of the male's motion and potential for producing that particular nose shape.

Moreover, not only are there motions and potentials by which the father conveys his particular morphological features such as nose shape, but there is also a potential and motion by which the male conveys his sex. Aristotle refers to such a potential and motion for producing a male offspring in the course of the discussion of resemblances in *GA* 4.3. There he says that while "what usually happens naturally is that [the motions] *qua* male and *qua* father master and are mastered together,"[32] this is not always the case. There can be mastery with respect to the male potential and motion, producing a male offspring, but also failure to gain mastery with respect to motions and potentials corresponding to the father's particular traits:

> If the motion drawn from the male gains mastery but that from Socrates does not, or the one from Socrates gains mastery but the one from the male does not, then it turns out that a male resembling his mother comes to be and a female resembling her father. (*GA* 4.3, 768a28–31)

Given the terminological similarity between the discussion in *GA* 4.1 and *GA* 4.3, and given the reference within the discussion of inherited traits to the motion that produces males, it is reasonable to assume that Aristotle

[31] According to Balme 1987a: 292, these motions and potentials are not additional but rather identical to the form passed from the father. There is an argument against this interpretation and for an alternative construal of the relation between the motions and substantial form in Gelber 2010:204–9 and n. 28, n. 59.

[32] *GA* 4.3, 768a21–22: "μάλιστα μὲν οὖν πέφυκεν ᾗ ἄρρεν καὶ ᾗ πατὴρ ἅμα κρατεῖν καὶ κρατεῖσθαι." I am assuming, as does Peck (1942) and Henry (2007: 258), that the unexpressed subject of κρατεῖν and κρατεῖσθαι is "motions," κινήσεις, from the preceding sentences.

thinks the same causal mechanism is involved in both sex determination and morphological resemblance.[33] It is reasonable, consequently, to take Aristotle's discussion of the causes of resemblances in *GA* 4.3 to be elucidating the details about sex determination omitted in *GA* 4.1.

There is no explicit mention of any motion or potential for conveying female sex in that discussion. There is, however, a reference to motions and potentials for maternal resemblance. Aristotle refers to these maternal motions and potentials in the course of explaining resemblances to remote ancestors. In short, there are two forms of failure that can occur at the level of these motions and potentials. The first type of failure is "displacement" (ἔκστασις), which involves a "change into the opposite." Displacement explains why there can be resemblance to maternal rather than paternal ancestors.[34] The other type of failure is "relapse" (λύσις), which explains how the offspring can resemble more remote ancestors on either the maternal or paternal side. When he discusses this second type of failure, it is clear that there are maternal motions and potentials as well.

> In this way [both on the side of the males and] also on the side of the females (θηλειῶν), [the motion], from the female generator (ἡ τῆς γεννώσης) [relapses] into [the motion] from her mother (εἰς τὴν τῆς μητρός), and if not into that [motion], then into that of her grandmother. (*GA* 4.3, 768a18–21)[35]

So, there are motions and potentials for resemblances to the mother and her ancestors. When the male's motions and potentials fail to master and displacement (ἔκστασις) occurs, corresponding motions for producing likenesses to the female in the respects in which the male potential failed to gain mastery become operative. Those motions and potentials then produce the features of the offspring by which it resembles the mother

[33] Alternatively, Henry takes the explanation of sex determination, which is ultimately a function of the degree of heat in the organism's heart, to be given in *GA* 4.1, and he takes the mechanism described in *GA* 4.3 as explaining inheritance of sexual organs (Henry 2006a: 271–2). Although Henry is right that Aristotle distinguishes sexual morphology from "functional" sex (the ability or inability to concoct semen), he gives no argument for there being separate causal mechanisms for these. In the absence of an argument, I am inclined to disagree. It seems to me more economical to assume there is only one mechanism being described, first only in rough outline in *GA* 4.1, and then in greater detail in *GA* 4.3.

[34] Cf. *GA* 4.3, 768a2; a4; a15; a26–27; b8; b25. There is far too little said about ἔκστασις to be certain about the details. It is not clear what, for instance, undergoes ἔκστασις or what changes into the opposite. Translators supply such different subjects as "movement," "material," and "male principle." In an attempt to remain neutral, I simply speak of the result being the opposite of what it would have been.

[35] I am taking the unexpressed verb here to be λύονται from the preceding sentence, at line a15, the subject of which is motion (κίνησις).

(or more remote maternal ancestors, depending on whether relapse occurs as well). Assuming that roughly the same account of the mechanism that produces familial resemblances is supposed to apply to the determination of the embryo's sex, we get the following result: the production of a female is due to the motion and potential for producing females becoming operative when the motion and potential for producing males does not master.

Assuming this is correct, the cause of sex determination is not merely the "interactions of material forces" operating in the way that "cold air acts on the surface of a pond forming ice."[36] For, while there is a role for heat and coldness in Aristotle's biological account of sex determination, that role is too remote. Aristotle, in fact, explicitly criticizes people such as Empedocles who claim that the cause of sex is heat and coldness, on the grounds that this is insufficient to explain how sexual organs are produced:

> Well then, to suppose the cause of the male and the female is heat and cold, and the secretion having come from the left or right, is somewhat reasonable . . . But to reason thus is to grasp the cause from too far away, and it is especially necessary to move toward it from as close to the primary causes as possible. (*GA* 4.1, 765a34–b1, b4–6)

The proximate, primary cause of the production of a female is the motion and potential which becomes operative when the male motion and potential does not master. To claim the cause is material necessity is, as Aristotle says, to "grasp the cause from too far away." The alternative reading of Aristotle's account is, for this reason, overly simplistic.

Further, the alternative account goes too far in its rejection of the view that generation aims to produce male offspring. For, by describing the process simply as a matter of "non-teleological necessity," the alternative account infelicitously depicts Aristotle's account too neutrally. Although the birth of a female is not a failed attempt at the production of a male, Aristotle's explanation does not treat male and female births equally.

First, male motions dominate when there is συμμετρία between the parents, whereas female births are described as occurring in cases where there is a lack of συμμετρία (*GA* 4.2, 767a22–27). Assuming that συμμετρία here means something such as "commensurability" or "due proportion," it would appear that females occur when there is a disproportional relationship. This is why female offspring occur when the parents are too old, too young, or sick (*GA* 4.2, 766b28ff, and 767a25–26; *GA* 4.3,

[36] Henry 2007: 257n18.

767b11–12). In such cases, there is a lack of proportion, perhaps less kinetic heat than is proportionate to the amount and temperature of the matter.

Second, whereas the motions that produce resemblances to the father are present "in actuality," those that produce maternal resemblances – or at least that which determines female sex – are said to be present "in potential" (*GA* 4.3, 768a11–14).[37] Although it is unclear what exactly it means for a motion to be present "in potential," the fact that it applies to the motions and potentials for maternal resemblance (and resemblance to ancestors) indicates that Aristotle is not thinking of the motions and potentials for paternal and maternal resemblance as being on a par.

Because the alternative account suggests something much more egalitarian, and since material necessity is not the primary cause of sex determination, the alternative account misdescribes Aristotle's view.

A Third Option

So far, I have argued that neither the standard reading nor the alternative account is completely accurate. The standard reading depicts the process of generation as though it is supposed to result in a male, and that females come to be accidentally. If that were so, females could not also be for the sake of something, as Aristotle thinks they are. A problem for Henry's alternative is that it suggests Aristotle's account is more neutral than it is. I agree with Henry that, considering animal generation at one level of generality, Aristotle thinks both males and females are successes. Strictly speaking, the aim of generation is an offspring the same in form as the parents. However, when Aristotle discusses the mechanism for conveying sex (and other resemblances), there is an asymmetry: Males are the results of a motion to produce males, which motion is present in actuality and which will master provided there is commensurability or due proportion (συμμετρία); females, on the other hand, result when there is not συμμετρία and the male motion fails to master, allowing the female motion and potential to become operative.

What is needed, therefore, is an interpretation that acknowledges that the female motions are only operative when the male motions and

[37] *Where* these motions and potentials are present is vexed, though it is not crucial for my purposes to adjudicate this. Cooper 1990: 70n14 claims these are present in the male semen. However, it is possible (and more plausible) that these are present either in the menses or in the κύημα (the first mixture of male and female principles), or even that these are potentially present in – i.e., "during" – the process of generation.

potentials fail without saddling Aristotle with the view that females are merely lucky accidents. As I read *GA* 4.1–3, there is a motion for producing females in Aristotle's causal account. So, at least there is some *per se* cause for the production of females. However, it is still not clear why it is not just a lucky accident, in Aristotle's view, that this motion to produce females ever does become operative, let alone that it does so with any regularity. How can Aristotle be thinking that female motions are operative in the absence of συμμετρία, and not therefore also think that females are failures or accidents?

An answer to this emerges by considering what the circumstances would have to be like for male motions to always be dominant. This would require, for each successful generation, perfect συμμετρία between the parents. Such συμμετρία would always obtain in conditions in which climactic factors, age, and health of the parents were fixed and stable. But Aristotle knows that those are not the conditions we are in. The sublunary realm is vagarious, and living beings are enmattered, and thus subject to changes of many kinds.[38] The contingencies of the sublunary world and the variations in the quality and quantity of matter are already built into Aristotle's causal explanation, and the causal process that explanation mirrors. Consequently, his explanation presupposes that male mastery will not always occur.

This is reinforced by a remark Aristotle makes in the course of explaining why females are produced:

> Since it is possible for the male not to master either because of youth or old age or some other such cause, the female offspring among those animals necessarily comes to be. (*GA* 4.3, 767b10–13)

Aristotle could be taken to be claiming here that it is possible for the male not to master, and *in those cases* a female necessarily comes to be. This is not likely, however, since according to Aristotle's account, it is not necessary that females come to be when the male principle does not master; it is possible in those cases for a monstrosity to result, or no offspring at all. Instead, I think Aristotle is saying that it is the fact that

[38] Aristotle is seldom explicit about this background assumption – i.e., the assumption that there is something indeterminate about how things will go for enmattered beings – though he occasionally mentions it. For example, at *GA* 4.10, 778a4–9, when discussing the connection between "cosmic" periods and the life cycles of living beings, he says the following: "Nature's aim, then, is to measure the generations and endings of things by the measures of these bodies, but she cannot bring this about exactly on account of the indeterminateness (ἀοριστία) of matter and the existence of a plurality of principles which impede the natural processes of generation and dissolution and so are often the causes of things occurring contrary to nature" (Peck translation, slightly modified).

it is possible for the male principle not to master – i.e., that it is not guaranteed or necessary that it always does – that explains why, necessarily, females sometimes result. In other words, it is the possibility of the male motion not mastering that makes female births necessary.

To be sure, the male motions and potentials, unlike the female's, are "present in actuality" and not merely in potential. So the male motions are, in a sense, privileged. In my view, what explains this privileging of the paternal motions is not the assumption that the whole process is aiming to produce males that perfectly resemble their fathers. Since exact replication almost never occurs, I do not think Aristotle could suppose that generation aims at that. Rather, since the male is the active factor in generation, it is more plausible to assume that all of the motions and potentials he provides are *ipso facto* the similarly "active" ones. That is, the male motions have this status solely in virtue of the privileged causal status that the male has, in general, *qua* active factor in the substantial change.[39]

This interpretation, moreover, allows us to take seriously Aristotle's claims that females are for the sake of something. There is no conflict, at any rate, between the idea that females are not produced when there is συμμετρία between the parents, and the idea that females are for the sake of something. For, in general, the fact that something will not occur under all conditions is surely compatible with that thing also being for the sake of something. The fact that plants do not grow leaves under any and all conditions, for instance, does not prevent leaves from being for the sake of something. And the fact that something will not occur under some specific conditions does not necessarily conflict with it also being for the sake of something, either. The fact that a plant does not grow leaves in the dead of winter does not entail that leaves are not for the sake of something. I can see no reason for thinking that the fact that females are not produced under conditions in which there is συμμετρία between the parents is incompatible with females also being for the sake of something. That is, the fact that males – and ones that bear perfect resemblances to the father – are produced when there is συμμετρία between the parents does not prohibit females from being for the sake of something. For, living organisms are not generated under conditions in which such συμμετρία always obtains. In the actual circumstances, Aristotle thinks, there are variations in the degree of heat and bulk of the

[39] Why it is that the male is the active factor is a further question. The answer to that question, I suspect, is that it was simply overdetermined. See Cook 1996 for a clear discussion of the various factors that would all be pulling Aristotle to this view of male and female reproductive roles.

spermatic residues.[40] Thus there is not always συμμετρία, and so Aristotle assumes that male motion's failure to master, and female motions becoming operative, is guaranteed to occur.

The fact that male motions will master when there is συμμετρία does not entail that animal generation is teleologically directed toward producing males. Interpreting the discussion of sexual determination in this way acknowledges that female births are due to a causal mechanism that has a secondary position, but does not require that females be failures.

Conclusion

I have argued that the standard reading of Aristotle's embryology, according to which females result from a defective teleological process, cannot be correct. There is no direct evidence for that reading, and it is difficult to reconcile with Aristotle's commitment to females being for the sake of something. The alternative account, according to which the cause of sex determination is merely material necessity, is also not correct. That account is too simplistic, and it glosses over the asymmetries in Aristotle's biological theory.

I proposed instead that we read Aristotle's explanation of the mechanism by which sex is determined as running parallel to his account of inherited characteristics. This mechanism is below the level of εἶδος transmission, which is the goal or aim of generation. At that lower level, the paternal motions are the ones that master when there is συμμετρία, simply because the male is the active factor and thus his motions and potentials are present "in actuality." But this is compatible with the production of females – which occurs when the female's motions become operative – also being for the sake of something, since natural generation does not take place under conditions in which there is always συμμετρία between the parents.

No doubt Aristotle thinks that females are inferior to males, and his society considers a man unfortunate who does not have children who resemble him. But if Aristotle thinks these views are grounded in biological facts, it is not that those nonresembling children and daughters are failures.

40 That there are such variations is all for the good of any animal kind, since it ensures that there will not always be συμμετρία. I will not argue for this here, but I suspect that it is no accident that there are such variations, for Aristotle. If so, this would lend some support for an interpretation of Aristotle's teleology as extending more globally of the sort that David Sedley (1991, 2007, 2010) has defended.

CHAPTER 10

Something(s) in the Way(s) He Moves
Reconsidering the Embryological Argument for Particular Forms in Aristotle

Gregory Salmieri

Introduction

As my subtitle indicates, this paper concerns the argument that Aristotle's embryology entails the existence of robustly particular forms (εἴδη). By "robustly" particular, I mean that each individual organism has a form that is distinct from that of other species-members not merely in number, but also by including or being responsible for some of the characteristics that distinguish the individual from (at least some) other species-members.[1] John Cooper puts the point as follows:

> Some have thought from reading just the *Metaphysics* that an Aristotelian form is a non-repeatable instance of some general specific type, differing from other instances not internally [. . .] but only by the accidental historical facts about the individual whose form it is by which we mark that individual off from others of the same species. The *GA* makes it clear that, on the contrary, each form [. . .] includes [. . .] all those distinctive characteristics of structure and organization for which in the individual whose form it is it is directly responsible [. . .]. Thus Aristotelian forms are particular in that each form contains within itself the basis for its differentiation from (as well, of course, as its affiliation with) other forms of the same specific type.[2]

[1] Of course, εἶδος often means species and is contrasted with genus (γένος). When the word is used in this sense, it is incoherent to say that two members of the same species differ in εἶδος. But εἶδος also functions in contrast with matter (ὕλη), and this is the sense in which it is usually translated "form." What Aristotle calls εἶδος in this sense is what makes an individual composite substance the substance it is, as opposed to being merely some material that has the potential to be that substance. Part of the substance's being the substance it is (as opposed to being any of the other alternatives admitted by its matter) is the substance's being a member of its species. The robustly particular forms view holds that, in addition to being responsible for this species-membership, the form is responsible for at least some of the characteristics that differentiate the individual substance in question from other members of its species.

[2] Cooper 1990: 83–4.

Those who argue that the *GA* makes this "quite clear" include Cooper (1990) and David Balme (1962, 1987a). The argument has been criticized by G. E. R. Lloyd (1990: 16–24) and Jessica Gelber (2010). Part of my purpose in this paper is to defend the argument, but my primary purpose here is to raise questions about what individuates forms, and so what it is for one of an organism's characteristics to be or not be part of its form. It is different answers to such questions, I will suggest, more than different interpretations of particular passages from the *GA*, that lead scholars to disagree about whether Aristotle's embryology commits him to the existence of robustly particular forms.

In the first section, below, I sketch out the embryological argument in general. I go on, in the next section, to consider differences between Balme's and Cooper's understandings of Aristotle's view of reproduction, and I give my reasons for preferring an interpretation nearer to Balme's. In the subsequent two sections, I consider the two main arguments that I find in Lloyd and Gelber. The second of these leads me to the question of how forms should be individuated, which I discuss in my final section.

The Embryological Argument and the Issue of Individuating Motions

The embryological argument can be summarized as follows: According to *GA* 1.22, a father's contribution to his offspring is a set of movements that carries his form and is conveyed to the offspring through the semen. All the matter is supplied by the mother, who also has a secondary influence on the offspring's form insofar as her own movements are present in the matter she supplies. Since the father contributes only form and no matter, any characteristics inherited from father to child must be due to the father's form. Since Aristotle recognizes (in *GA* 4.3) that a child can inherit the father's distinctive characteristics, he must have understood these characteristics to be part of the father's form. Therefore, the form that a father passes on to his child must be a *robustly particular* form, distinctive to him, rather than a universal human form. And indeed, in *GA* 4.3, Aristotle speaks of movements that come from the particular fathers insofar as they are the particular men they are.

Here are the two key passages:

> One may also grasp from these examples how the male contributes to generation. For not every male emits semen, and for those males who do emit, it is no part of the fetus that is produced, just as nothing travels from the carpenter into the matter of the timber, nor is any part of carpentry in

> the product; but, rather, a shape (μορφή) and form (εἶδος) are imbued from this through the movement in the matter. And, while the soul (in which is the form and the knowledge) moves the hands or some other part in a certain sort of motion (a different [motion] when the product is different), the hands move the tools; and the tools, the matter. Likewise, the nature in the male (among the semen emitters) uses the semen as a tool and bearer of a motion in actuality, just as in artistic productions the tools are in motion (for the art's movement is somehow in them). Those animals that emit semen contribute in this way to generation, but those that do not emit and in which the female inserts some part of herself into the male, are like when someone brings matter to the craftsman. For due to the weakness of such males, their nature cannot produce through other means; but even when it applies itself direct, its movements are barely sufficient; it acts like modelers rather than carpenters; for it crafts the creation not through something else but directly by using its own parts. (*GA* 1.22, 730b8–32)[3]

> When [the male's motion] masters, it produces a male and not a female and a resemblance to the generator rather than to the mother; whereas, when it does not master, with respect to whatever potential (δύναμις) it does not master, it produces a defect in that [potential]. By "each potential," I mean: the generator is not only male but also such a male – e.g., Coriscus or Socrates – and not only Coriscus, but also a human being. And it is in this way that among the [items] belonging to the generator with respect to being a generator (rather than belonging incidentally – e.g., the generator is literate or someone's neighbor): some are nearer and some farther. What is more distinctive and particular is always stronger relative to generation. For Coriscus is both a human being and an animal, but human being is nearer to distinctive than animal is. And both the particular and the kind reproduce, but more so the particular; for this is the substance. For, though the offspring also comes to be a certain quality, yet it also comes to be a certain this – and this is the substance. Therefore, the movements in the semen are from potentials such as all of these, and in potentiality also those of the ancestor, but those of some particular are always much closer – by "a particular," I mean Coriscus or Socrates. (*GA* 4.3, 767b20–768a2)

The motion that emanates from the father according to the *GA* 4.3 passage and that corresponds to his various potentials would seem to be the very form-conveying motion discussed in *GA* 1.22. The form is *something in the way he moves* the semen and, through the semen, the matter in the mother's uterus. Thus, my title. However, whereas *GA* 1.22 and the beginning of the *GA* 4.3 passage speak of this as a single motion, the second passage goes on to speak of multiple motions corresponding to the father's multiple potentials (767b36). It is not obvious whether

[3] Translation liberally adapted from Balme 1992.

there is meant to be one motion or several. Thus the ambiguities in my title: *something(s) in the way(s) he moves.*

The proponents of the embryological argument see the seminal motion as one complex motion transmitting one complex item – the form – from father to child. This form then is responsible for the child's being a substance like his father, where this likeness includes both the child's membership in the father's species and any other inherited likenesses that may distinguish father and child from other species-members. The critics of the embryological argument deny this. In effect, their criticism amounts to raising and defending the alternative view that it is by separate transactions, all occurring through the medium of seminal motion, that the father causes the child to be a member of his species and also to have other inherited traits.

Differences between Balme and Cooper

There are some differences between Balme's position and Cooper's concerning which characteristics are conveyed from father to child via seminal motion, and are (therefore) part of the father's form. Aristotle makes clear at 767b28 that such accidental characteristics as being literate or someone's neighbor are not included; he says these characteristics do not belong to the father *as a generator* (καθὸ γεννητικόν). Which traits, then, do belong to the father as a generator? Balme (1990: 51) thinks this category includes at least the full range of traits that we today regard as hereditary. Cooper (1990: 81–84; cf., Gelber 2010: notes 17–18) includes the father's "essential organs and limbs, and all their properties (even the individual ones) that are relevant to their correct functioning," but he excludes "biologically superficial characteristics like eye color, pitch of voice, and coarseness or fineness, length, quantity and natural straightness or curliness of the hair," on the grounds that Aristotle gives material causal explanations for these traits in *GA* 5.1–4.

I think the evidence favors Balme's position (or perhaps a moderated version thereof) over Cooper's. *GA* 4.3 gives us the treatise's only account of how features can be inherited from the father, and *GA* 1 strongly suggests that the phenomena Aristotle thinks a theory reproduction must explain include the heritability of some of the paternal traits that Cooper excludes:

> These are roughly the main grounds for believing that the semen comes from all the body. But if we test the argument, we come rather to the opposite conclusion. For it is not difficult to refute what has been said.

> First, to show that resemblance is no indication that semen comes from all the body there is the fact that resemblance occurs in voice, nails, and hair, and movement, but from these nothing comes away. And there are some things that men do not have by the time they generate, for example gray hair or a beard. Further, there are resemblance to distant ancestors from whom nothing [i.e. no semen] has come; these resemblances occur at intervals of several generations, for example the woman at Elis who had intercourse with the Ethiopian – it was not her but her daughter's son that became Ethiopian. (*GA* 1.18 721b35–22a11)

Cooper (1990: 80) thinks that the dialectical context of such passages in *GA* 1 precludes us from inferring from them which resemblances Aristotle thinks are hereditary and so seeks to explain by his theory of reproduction. But Aristotle's point in citing the fact that children can resemble parents in beards, baldness, and in the characteristics (presumably of skin and hair) that make one "an Ethiopian" is surely not only that children resemble parents in features from which seed cannot come. If this was the point, Aristotle may as well have pointed to resemblances in what language the people speak, or in what books they have read.[4] His point is that the resemblances that a theory of reproduction needs to explain include some that cannot be explained by the theory that seed comes from all parts.

Cooper's primary reason for excluding traits like the color of eyes, skin, hair and the like from the form seems to be the material causal explanations given of these traits in *GA* 5, but at the beginning of that book Aristotle tells us that the he is explaining in terms of material *and efficient* causes (778b1), rather than in terms of final causes and essences. So, if we think that, in transferring his form (including any features of this form that distinguish him from other members of his species), a father is functioning as an efficient cause, then it is easy to see in principle how this form-transference can find a place in the explanations of *GA* 5. And if we look at *GA* 5's explanations in particular, it is easy to see how they are compatible. *GA* 5.1, 779b14–780a24, tells us that eyes are dark or light according to how much fluid they contain. Cooper (82) says that the amount of fluid is just an effect of how much happens to be available during reproduction.

[4] Aristotle does refer later to the fact that fathers and sons often also resemble each other in wearing the same type of shoes and he notes that this does not imply that semen comes from the shoes (732b31), but here the context has already shifted to the issue of the rival theory's failure to notice that two things can resemble because their efficient causes are the same or similar (as in the case of two shoes made by the same cobbler), rather than because an influence has propagated from one resembling item to the other. Thus resemblance between a part of a father's body and a corresponding part in his son's need not be propagated from the father's part to the son's – both can be a products of the same nature. And, of course, Aristotle thinks that this nature is propagated from father to son.

But the text he cites (778a35ff) doesn't say this, and it is hard to see how this could be Aristotle's view, since he thinks that human eyes are all blue in infancy and later differentiate into different colors (779a26–28). It is more likely that how much fluid there is has to do with the space for fluid, and that this is determined largely by the form. Since infants are so small, even those with the proportionately largest spaces for eye fluid will have small spaces by adult standards, and so they will have blue eyes.

Moreover, though Aristotle thinks that green eyes are overall best, there are indications that he thinks each eye color has something to be said for it. The shallowness that makes blue eyes too sensitive to see well in the day and the depth that makes brown eyes not sensitive enough at night (780a1–6) would presumably give blue eyes an advantage over green in the night and would give brown eyes an advantage in unusually bright environments. Though Aristotle doesn't say this, it is consistent with his view that people differ in eye color in so as to be optimized for seeing in different conditions. This could be so, even though green eyes are best *simpliciter*, because they enable one to see well see in the widest variety of conditions.

Even if (or in those cases in which) the eye color is determined by how much fluid is available for eye construction, how much is available is caused not only by material factors, such as how much the mother has had to drink and which winds are blowing, but also by how the materials are apportioned by the efficient cause in the process of generation. At any rate, this is how it works in the arts: a baker making a cake out of a set of ingredients that includes less than the ideal amount of water would have to decide how to deal with this deficit; some bakers may make the dough less than optimally moist, whereas others may instead skimp on the water in the icing. Since the father's influence is analogous to that of a craftsman we should expect the form he contributes to make such analogous determinations in the case of reproduction with suboptimal materials. If so, then a son's having blue eyes like his father will be caused in part by the father's having transmitted a form that deals with insufficiently moist matter by economizing on the eye fluid rather than elsewhere.

(If we conclude that Aristotelian eye color determination works as described in this last paragraph, rather than as suggested in the previous two, then we are left with a moderated version of Balme's position. Balme held that the offspring of a fully successful reproduction would be an exact replica of the father, whereas on the present view, the blue-eyed father's reproduction would have been at least as successful had the child's eyes been green.)

Objection 1: The Mother Is a Cause of the Same Traits as the Father Causes

I turn now to the objections raised by Lloyd and Gelber. There are two objections that are raised (in different forms) by each of them. I begin with the one that I find less persuasive. It takes off from the observation that Aristotle recognizes that a child can inherit from the mother the same sorts of features it can inherit from the father. If we say that all the inherited traits are due to form, Lloyd and Gelber reason, it follows that the female contributes form as well, but this cannot be right, because it is an Aristotelian dictate and a premise of Balme's argument that the female contributes only matter and the male, only form.[5] I see no problem, however, in maintaining that the very same movements are a material cause when due to the mother and an example of formal or efficient causation when due to the father. Wouldn't we want to say something analogous in nonbiological cases in which hylomorphic compounds come to be?

Consider the case of stamping a signet ring onto a previously stamped piece of wax. The ring may fail to erase all signs of the previous stamp, so that the new seal has some features of the old. These features from the earlier seal will be part of the new seal's form in a perfectly straightforward

[5] Here are two statements of the objection: "Note, to begin with, a puzzling feature of this argument, which is that it starts from the firm distinction between form and matter and the insistence that the father's contribution to the offspring is form and not matter. But by the time we get to the conclusion it appears that the complete formal description is taken actually to include material accidents, such as sex and color. (On the female side of the equation, Balme explicitly notes that she does not contribute just matter, but also some formal characteristics: indeed all differences are for Balme formal differences and so any contribution from the female parent that helps to differentiate the offspring will be set on the formal side of the form/matter dichotomy.) On the view that all differences are formal differences, the characteristics that the offspring owes to the male parent belong, indeed, to form: but then it should not be deemed to be matter – and so the idea that the form includes matter (even in the shape of 'material accidents') falls to the ground. But of course it may be that the weak point in the argument as thus set out lies not in the conclusion, so much as in that assumption that all differences are formal ones. While one can see that in a sense to differentiate is to pick out some determinate character, and given that matter is in itself indeterminate, differentiation must be through form, there is another sense in which it is perfectly possible and indeed desirable to speak of differences in the matter of two individuals of the same kind. Aristotle still needs to distinguish the way in which a bronze triangle differs from a wooden one from the way in which a bronze triangle differs from a bronze statue" (Lloyd 1990: 17). "However, the offspring will resemble the mother when the male's κινήσεις fail to prevail, and so the κινήσεις that she provides are used during the formation of the embryo. And as Aristotle says again and again, what the mother provides in animal reproduction is matter and not form. But if the subspecific, inherited features that these κινήσεις produce are part of form, then either she also contributes form (and so despite what Aristotle initially says, the mother must end up making the same sort of contribution as the male does), or she does not really contribute κινήσεις for subspecific features" (Gelber 2010: 197).

sense: they will be literally part of the seal's *shape*, and, if an impression were to be made of the new seal (e.g., by filling it with clay), these features would be passed on along with those imprinted from the second signet ring. However, if we consider the transaction by which this second seal comes to be, it is clear that the second ring alone transmits form in this change. The earlier ring's shape, which is already present in the matter, is a contrary of the form being transmitted, and if any remnants of this earlier shape remain in the finished seal, it is due to a failure of the new form to displace its contrary and thereby master the matter.

For a second example, consider a beam, which comes to be a lintel when it is raised into the relevant position and held up there by supports. The beam is matter for the lintel, and the form is the position imposed by the builder. Since the beam's is an earthen object, its natural motion is to go down. Suppose that, because of the builder's insufficient skills or tools, he is unable to fully overcome this natural motion, and the lintel droops in a way he did not intend. The resulting configuration will be part of the shape of the structure, but this is not because the beam contributed form. The beam is only a material cause, but because the material has a nature, it can act in ways that are contrary to the builder's attempt to impose form, so that the finished structure realizes only imperfectly the form intended by the builder.

As I interpret both of these examples, the matter, functioning only as material cause, nonetheless has an influence on the form of the compound. Likewise, I do not see why the menses contributed by the mother, which contains her motions, cannot contribute features that are part of the form of the offspring, even though these features are not part of the offspring's formal cause, and are not imposed by its efficient cause. Alternately, if we cannot consider these features as part of the form of the offspring unless we see them as having been in some sense aimed at by the efficient cause, we can adopt Cooper's explanation. On this view, when the movements from the semen fail to overcome the movements in the menses, they are converted into movements like those of the menses, and they then serve as an efficient cause of this form's being present in the compound.[6]

To see how this might work, consider how a sculptor's craft might be altered by his material. Classical, Renaissance, and Enlightenment

[6] By the time the motion undergoes this conversion, it may no longer be in the semen, but in the body (heart, etc.) of the forming embryo, but it is still the seminal motion in so far as (i) it was conveyed into the embryo from the father, and (ii) the principle by which the motion admits of alternation in light of its circumstances is part of this conveyed motion.

sculptors tried to give marble the fine texture/finish of their subjects – i.e., they wanted it to have the apparent texture of skin or lace, etc. Only if their skill was inadequate, or if their tools broke or they were interrupted, would their works retain an unfinished, rocky texture. Later sculptors like Rodin became intrigued with the raw texture of stone and started intentionally leaving that texture in their sculptures. In this case, the art was affected by the stone, with the result that the art now aims at rocky-textured products.

In this example, we have seen two ways in which the product of sculpture might end up with a rocky texture: either the art simply gives out and fails to alter the stone's texture, or the art is itself influenced by the texture of the stone and aims at including it in the final product. An analog to either alternative could occur in reproduction. Either the father's causal influence simply runs out at some point with the result that elements of an older form already present in the matter prevail *qua* material features (though they are features of the mother's form, their presence in the offspring is an instance of material causation) or else these features alter the efficient paternal motions, which thereafter imposes an altered form.

Objection 2: Efficient vs. Formal Causation

I turn now to an objection that I think is stronger. It can be summarized as follows: The mechanism described in *GA* 4.3 is one of efficient causation and efficient causation is a wider phenomenon than form-transference, so we need not understand everything that propagates from the father to the offspring as part of the form transferred by the father. Here is how Lloyd formulates it:

> It is tempting to assume that the efficient cause always just is the form in action, in its dynamic function, that is it always just transfers or imposes form. But we should be careful: the relationships between the form of the object produced (itself a compound – σύνολον – of form and matter) and the form of that which produced it (where again we may be dealing with an element in a σύνολον, a composite object, though sometimes it is just the form that acts) can be varied and complex.
>
> In some cases a proximate moving cause acts by transmitting its own form (shape) directly to the matter: a signet ring leaves its impression in wax. But in other accounts of the action of moving causes the situation is more complicated (21).

Sexual reproduction is more complicated because the father transmits his form not directly (as a signet ring transmits its form into wax), but via means that Aristotle analogizes to tools or to the role of heat in cooking.

Lloyd considers these means "proximate efficient causes" and he notes that they need not have the form of the product:

> [W]hat makes the bed (in the sense of the proximate moving cause) is the carpenter's tools, and (unlike the signet ring case) they do not themselves have the shape of the bed.
>
> [I]n these [cooking] cases too the proximate efficient causes do not themselves have the form eventually exemplified in the end product. The heat that cooks the eggs, sugar etc. into a soufflé is not itself a soufflé, nor even soufflé-like (22).

Thus, Lloyd concludes, the form transmitted by the father could be a universal human form, and the more specific resemblance to the father can be accounted for by effects of the bodily means.

I count two key ideas here: (1) the father's role as efficient cause not limited to form-transmission and (2) the specific motions (heating, etc.) can be the proper causes of the features by which the offspring resemble the father, even if these features are not parts of his form. Gelber makes these points as well, and she too uses the art analogy, arguing that, when an artist aims at a general end, his product acquires differentiating features as a result of something that is causally downstream from the art itself.

I doubt the claim that the artist's end is general. If there are a range of healthy bodily constitutions or styles of chair, then how can it be that, when a doctor or carpenter practices his art on a particular occasion, his aim is health-in-general or chair-in-general? Surely the difference between a Windsor chair and a Ming comes from the carpenter and not the lathe, and subtler differences too are also often aimed at by carpenters in specific cases. Nevertheless, Gelber and Lloyd are certainly right that sometimes, an artist causes effects in his products that are accidental to the end at which he aims.

To illustrate this Gelber gives the example of a French teacher who, by being bearded, influences the student to grow a beard. The student's beard is an accidental effect, not something he produces as a teacher, even though it is during the course of the French lessons that the influence occurs. What though, of the teacher's Parisian accent, which the student also picks up? Gelber argues that (unlike the beard) it is something that he produces in his student through his activity as a French teacher, but that it is distinct from the form (viz. French) that his teaching conveys.

She notes that we could say instead that the form the teacher conveys is specifically Parisian French (197), and she acknowledges that we could likewise say that a father conveys a "subspecific" form. But, she argues, we do not have to say this, and she thinks that, if we say it, we will need to contend with Objection 1 (discussion above), which she finds compelling.

Thus Gelber's preferred solution is that the motions in the semen are not a form being conveyed but the tools by which a form is conveyed, and these tools have effects other than conveying form. The movements in the mother are also tools that can be used to impose form.

Her evidence for this is *GA* 2.4, 740b25–34, which she takes to say that "the κινήσεις ('in' heat and cold) are used by nutritive soul in a manner analogous to the way that tools are used in craft production." But the passage does not say this. In Platt's translation, which Gelber uses, it reads as follows:

> [. . .] as the products of art are made by means of the tools of the artist, or to put it more truly by means of their movement, and this is the activity of the art, and the art is the form of what is made in something else, so is it with the power of nutritive soul. As later on in the case of mature animals and plants this soul causes growth from the nutriment, using heat and cold as its tools (for in these is the movement of the soul), and each thing comes into being in accordance with a certain formula, so also from the beginning does it form the product of nature. (*GA* 2.4, 740b25–34)

What is analogized to an artist's tools is not the *movements* in hot and cold, but hot and cold themselves. The movements of hot and cold are analogized to the artist's movements of his tools. And this movement is the *activity* of the art. I see no grounds here for thinking that this activity is anything intermediate between form and its effects, rather than *form* itself. Likewise in the case of nature, the motions in the hot and cold are *uses* of tools, and these uses are the activity of the form. If this is right, then, contra Lloyd, insofar as the hot and cold are like tools by which the father reproduces, they are not intermediate efficient causes between the father's form and its effect.

Heat is not the efficient cause of a soufflé. It is just the efficient cause of the insipient soufflé's becoming hot. And it causes this by being hot itself. The heat only makes a soufflé when it is judiciously applied as part of a complex process, and the efficient cause of the soufflé is the baker who orchestrates this process. The baking, including the heating, is the activity of the baker – an activity that he is able to perform insofar as his soul contains the form of the soufflé.

To be an efficient cause is to be a form-donor. If something did not donate the form of X, it is not X's efficient cause, though it may be the efficient cause of some feature of X or of something needed to produce X. Thus, heat – or a hot oven – may be the efficient cause of a soufflé's hotness, but not of the soufflé itself, even though the soufflé needs to have been made hot in order for it to come into existence.

If this is the right way to understand efficient causes, and a father is the efficient cause of his son's snub nose, this can only be because he possessed a snub-nosed form that he passed on to the child. Likewise, if the French teacher is the efficient cause of his student's Parisian accent, he must have a form of this accent that he transmitted through his teaching.

But, with Gelber, we can distinguish between the teacher's passing on both a Parisian accent and the French language, and his passing on Parisian French. Likewise, we can distinguish between the father's passing on a snub-nosed form *along with* a human form, and his passing on a snub-nosed, human form. This brings me back to the way I formulated the objection at the beginning of this paper. Granted that some form transmitted from a father to his child is responsible for each characteristic of his that the child inherits, why should we think that there is a single form – the form that makes the father the substance he is – that is responsible for all of these characteristics? Why can't the father, instead, pass on a collection of forms – one that makes him a human being, another that makes him snub-nosed, a third that makes him have a cleft chin, and so on?

On this view, each man possesses, in addition to the form that makes him the substance he is, any number of accidental forms, which he possesses not *qua* being a certain substance but *qua* being made of a certain parcel of matter. These other forms are describable as forms, and they are the sort of thing that can be conveyed to offspring through seminal motions, but they are no part of the form that makes him what he is, so that form can after all be a universal human form shared by all human beings. Posing the objection to the embryological argument this way reveals a deeper question: How do we individuate forms? What makes a form unified?

Individuating forms

Notice that the issue of individuating forms arises even in the simplest cases of form imposition – for example, the case of the wax and the signet ring. Suppose that the symbol on the ring is the letter F. The letter will be printed in some font – for example, Roman or Helvetica. If it is Roman, the F will have serifs. When the ring is impressed in the wax, and the shape of an F with a serif is produced there, is a single form of a serifed F conveyed, or are two separate forms conveyed, one of the F and another of its serifs? Surely here the right thing to say is that it is *one* form insofar as the shape of the serifed F acts as *one thing* on the wax. This is the right answer, even though the serifs are a comparative detail that might fail to show up in the impression if the wax is too hard or too soft, or if the ring isn't pressed down with sufficient force.

But, though we say that there is a single form that the ring stamps on the wax, there may nonetheless be perspectives from which we might regard it as multiple forms (e.g., as an F and as a particular font). We can see why if we think not of the impression that a ring makes on wax but of the impression made by a sheet of type on a page of paper when printing a page of poetry. If we focus on the process of the plate leaving an imprint on the paper, it will be natural to say that one form is conveyed, for the same reason that it is natural to say this in the case of the signet ring; but if we focus on the content printed on the page, we may judge differently. For example, if the page contains several, unrelated poems, we may think of each as a separate form simultaneously conveyed. Moreover, if we think of the content *as content*, we may still individuate it differently for different purposes. Each poem may have multiple stanzas, each of which has multiple lines; moreover, it may have multiple themes, which may be more or less well integrated into a unity. Likewise, when we think of the father's contribution to his offspring and try to individuate the forms conveyed, we can focus on the act of form-transmission or on the content of what is transmitted, and when we focus on the content, there are different purposes for which we might individuate it differently.

When Aristotle is thinking of the father as the efficient cause of the child, his language suggests that he is viewing the father as conveying a single thing – a form that makes the child a human being and one who looks more or less like the father. Thus *GA* 1.22 speaks of the form that the father contributes and *GA* 4.3 767b20 speaks of what happens when the male's motion masters. In both cases, the father's contribution is described in the singular. In *GA* 4.3, he begins to speak of *multiple movements* corresponding to *different powers*. This is to account for how the child could resemble the father in some respect while failing to resemble him on others. But even here Aristotle's treatment emphasizes the unity of these movements and powers rather than their distinctness. For example:

> So it's especially natural for [the motion or potential] due to maleness and [the motion or potential] due to the father to master or be mastered simultaneously; for the difference is small, so it's no task for both to result simultaneously – for Socrates is a man of a certain sort. That's why, for the most part, the males are like their father and females like their mother – for

> [in this last case]: displacement came about simultaneously for both [the male way and the paternal way]; and, while female is opposite to male, the mother is opposite to the father; and displacement is into the opposites. But if, while the movement from the male masters, the [movement] from Socrates does not master, or if the latter [movement] does [master], while the former doesn't, then it results that males come to be resembling the mother, and females resembling the father. (*GA* 4.3, 768a22–27)

> The patient displaces [the affecting thing] and is not mastered, either through a deficiency of the potency of the concocter and mover or through the abundance and coldness of what is being concocted and defined; for, while the mover, mastering here but not there, makes the conception multi-shaped, as happens in the case of athletes through their overeating – for due to the quantity of the food, the nature is not able to master it so that it grows proportionately and remains the same shape; the parts become other, sometimes almost so much that none looks like it did before. (*GA* 4.3, 768b25–32)

Notice that, in the first passage, where the movements that make the offspring resemble Socrates are distinguished from the movements that make the offspring male, Aristotle treats the resemblance to Socrates as a unitary thing, rather than as a manifold of independent characteristics.

In *GA* 4.3, movements (or potentials) are only distinguished from one another when doing so is needed to explain how the offspring can resemble in one respect but not another. The *default* mode is to treat the resemblance as a single whole, which might happen to get broken up in the struggle to master the matter, rather than as an *ad hoc* grouping that so happens to be traveling together. This interpretation is supported by the fact that Aristotle finds it necessary, in the second passage quoted above, to provide a special explanation for the fact that a form can master in one part but not another. There would be no need for this if he was not thinking of the likeness to the father as a unity (as a single form). This suggests that when Aristotle is focused on the father's role as efficient cause, he views what the father passes on as a single form, though it can be analyzed into component powers and though it is often realized incompletely such that only some of these powers are reproduced in the offspring.

However, there is another important context in which Aristotle (and we) might regard this same form as a motley collection made up of a generic human form and the forms for the father's various distinguishing characteristics. We can ask why a human being is a single thing, rather than a collection of parts and attributes, or why French is a single language, rather than a mere collection of words and grammatical rules uttered and used in various ways by various people. The accounts that reveal the unity of these phenomena operate at a level of generality that doesn't specify many of the determinate details of individual members of biological species or of the idiolects and accents of individual French speakers.

Accounts of why an animal's many parts and attributes constitute a unity can be given at various levels of generality, but there is a narrowest level after which such accounts can no longer be given. To use an example first offered by James Lennox, consider how the features shared by birds (wings, beaks, talons, etc.) combine to make up a distinctive type of animal adapted to a distinctive way of life. One can account for this unity at the general level of birds as such, but one can also give more specific accounts, for broad kinds of birds, and for species within those broad kinds. This is because the various respects in which species of bird differ from one another (e.g., the lengths of their legs, the lengths of their necks, and the shapes of their beaks) are not all independent. It is no accident, for example, that swamp-dwelling birds, which have long legs, have correspondingly long necks and broad beaks. The birds need the long legs to wade through their environment; and, given the length of their legs, they need long necks to reach the ground, where they use their broad beaks to dig food out of the mud.[7] Thus, the several features, along different axes, that distinguish one bird species from the others in its kind constitute (as Lennox puts it) "a fully coordinate set of structures and activities" suited to its particular life and environment.[8] Contrast this with an individual crane. His neck may be slightly longer than the necks of some other cranes, and slightly shorter than the necks of others, likewise with the width of his beak. While there is a teleological explanation for the co-occurence of crane-range leg-lengths with crane-range beak-widths, there is no such explanation for why a neck of this absolutely determinate length should be found in a bird with a beak of this absolutely determinate width. The co-presence of any pair of fully determinate characteristics is due either to efficient cause (if the both characteristics were inherited from the father) or a material cause (e.g., if the characteristics were both inherited from the mother or if they resulted from material conditions that obtained during gestation), or from the accidental features of the interplay between the material and efficient cause (if the characteristics are inherited from different parents).[9] If the neck-length and beak-width

[7] *PA* 4.12, 693a15–17 and 694b12–16. See Lennox's discussions in 1987: 356 and 1985: 80–1.

[8] Lennox 1987: 358.

[9] It follows that it is possible to give explanation of such determinate characteristics, along the lines given in *GA* 5, but these will not be explanations of why some *kind of animal* should possess the relevant characteristic, but rather why *any animal* should possess it, if its matter happens to be constituted in a certain way.

are heritable, then they constitute part of the particular crane's form that he can pass on to his child, but differences at this level of specificity between him and other cranes do not constitute otherness-in-form because they are teleologically irrelevant.[10]

Lennox explains:

> Every individual of a kind will have its organs and tissues completely differentiated – they will have a precise hardness or softness, texture, hue, temperature, viscosity, width, length, and so on. But there will only be a *range* of these precisely determined qualities relevant to the organism's life. Individuals which differ *within* this range do not differ essentially or in form, but only "incidentally." Two individuals may be one in form while having every qualitative property instantiated to a different degree within the appropriate range. But these individuating differences will not be relevant to a functional account of the organism's nature [...]. If to be one in form is to be indivisible in account in the way specified, then numerically many individuals will be indistinguishable with respect to that account. If that account names only teleological features, those individuals will be identical with respect to those features required for each of them to exist. Those features, however, will vary from individual to individual in ways which do not entail altering their functions in ways crucial to the organism's life. Such variation between individuals is compatible with their being one in form. (1985: 81–2)

To support this view of oneness in form, both Balme and Lennox point to *GA* 5.1, which limits teleological explanation to (determinable) features common to all the members of species:

> About these [viz. eye color, height, voice pitch and features such as hair or feathers] and such things, we must no longer think that there is the same manner of cause. For whichever features are neither common to nature nor distinctive of the particular kind, none of these either are or are generated for the sake of anything. For, while the eye is for the sake of something, [its being] blue is not for the sake of anything, unless the affection is a distinguishing characteristic of the kind. In some [animals], it is not connected with the account of the substance; but, since it comes to be by necessity, one must reduce [it] to the matter and the moving principle. (*GA* 5.1, 778a30–36)
>
> Thus each thing is for the sake of something; through this cause and through the remaining [causes] everything is generated that either belongs in the account of a particular thing, or is what the thing is for the sake of, or else is for the sake of something. And, whatever generation there is amongst

[10] Balme 1987b: 298.

> things that are not such, the cause of such things must be sought in their movement and in their generation, since the differences occur in the very composition [of the animal]. For, while it has an eye from necessity (for an animal is supposed to be such), though such an eye is from necessity, it is not from the same necessity but another manner [of necessity]: that it is such as to act such by nature or to be affected. (*GA* 5.1, 778b10–19)

The precise way that the determinable characteristics of a given species are determined in any individual member of the species doesn't have a final cause, and Aristotle is telling us that it needs to be explained by reference to material and efficient causes – that is, by reference to the matter provided by the mother and by reference to the father's role as efficient cause. I have been arguing that this role consists in the father's transmitting to the child a single form that includes all his heritable features (both those that are determinable and characteristic of the species, and those that are determinate and distinctive to the father). The fact that Socrates' skin, nose, and hair are all of the determinable human type can be explained final causally, in terms of the role that such features play in a human life; and this same fact can be explained efficient causally as the result of the form impressed by Socrates' father on suitable matter provided by his mother. But the fact that Socrates has specifically *white* skin, *wavy* hair and a *snub* nose can only be explained in the latter of these two ways. This is because this specific combination of determinate traits is an accidental unity.

Socrates has the combination, not because they belong together, but either because they were all part of the form impressed on him by his father – or because they were all part of the residual maternal form present in the matter – or, if some characteristics came from each parent, because of the intricacies of the way in which the paternal form mastered and failed to master the matter. This combination is part of Socrates' determinate form, which he will pass on to his children. If one of his sons fully realizes this form, and comes out exactly like Socrates, it won't be an accident that the son is like the father, but the precise combination of traits in which they are alike will still be an accidental combination – one with no teleological unity. In any case, reproductive process has a limited resolution that makes this outcome unlikely: the determinable combination that is distinctive of the human species is reliably reproduced from generation to generation, but the highly determinate combination of characteristics that is distinctive to Socrates is not.

Moreover, since there is nothing *good* about the distinctive Socratic combination of determinate traits other than their falling within the determinable human ranges, the limited resolution of the reproductive process is hardly a *failing*. So long as the child is a functional human, the process was a success. Because of this we can say that the goal of reproduction is to reproduce the determinable species form, rather than the robustly particular form of the father.

All the same, I can see no reason to think that the determinable species form and the determinate characteristics of a father are not separate items in the father that are conveyed separately to his offspring. This would mean that there was one form in Socrates that caused his nose to fall within the human range of shapes, and then a separate one that caused it to have the specific snub shape it does, and that Socrates then passes both of these forms (and a host of others) side-by-side to his children. I know of nothing in the text that positively rules out such a view, but there is also nothing to suggest it. And the texts cited above in support of the view that Aristotle thinks of the father's form as a single item suggest against it. Moreover, the view strikes me as odd in its own right. It is more natural and more in keeping with the spirit of the *GA* to think that Socrates causes the child have a human form precisely by causing him to have a form like his own, even if the resolution of the process by which he does this is such that the child will almost always be human, usually resemble Socrates in particular, and hardly ever be a clone.

This brings me back to Gelber's French teacher. It is a false alternative to say that his lessons either convey his Parisian accent as something separate from the French he teaches, or that they are lessons in Parisian French. One cannot pronounce French at all without pronouncing it in some accent, and so to teach a student to pronounce French is to teach the student some specific pronunciation in some accent. For most purposes, which accent it is doesn't matter so long as it falls within the acceptable French range, so we think of the teacher as a French teacher, rather than as a Parisian French teacher, and we think of him as having been successful so long as the student ends up speaking acceptable French. To be a speaker of a language is to have some specific ability to speak it consisting in a specific idiolect and a specific accent, and to teach the language is to convey this ability to the student; the aim isn't necessarily that the student speaks just as the teacher does, for a range of idiolects and accents are equally acceptable, but there is no generic ability to speak (absent any idiolect and accent) for a teacher to convey.

Likewise, I think, to have the human form is to have some robustly particular human form; and to father a child is to transmit this robustly particular form to him – via a process that hardly ever reproduces the form exactly, but reliably reproduces it with enough detail to preserve everything that contributes to the human way of life. So long as this is preserved, the child's form is identical to the father's at the lowest level of detail at which there is any teleological unity to the features that make up a human being. Thus, there is a sense in which it is right to say that the child fully has the father's form, that it is a generic form shared by all human beings, and that any more determinate features shared by father and child are accidents due to the father's role as efficient cause.

If I am correct, then to speak in this way is to consider the form at the level of resolution that brings out its teleological unity, and at which the form is reliably preserved in reproduction over the course of generations. But the details we abstract away from when we view the form in this way are not due to any separate factor in the father, and the way in which the father causes his children to resemble him in these details (when they do indeed resemble him) is no different from the way in which he causes them to resemble him in being a human. He produces both sorts of resemblances by the single act of transmitting one form that is robustly particular, though we must abstract away from much of its particularity to appreciate certain aspects of its causal role.

I consider it an asset of this interpretation that, while maintaining that an individual's form is robustly particular, it allows for a significant sense in which this same form qualifies as the generic human form. For, given the well-known conflicting texts on the universality or particularity of forms (especially in the *Metaphysics*), we should expect there to be something right in each of the contrary answers.

CHAPTER 11

Aristotle's Explanations of Monstrous Births and Deformities in Generation of Animals *4.4*

Sophia Connell

Introduction

In *GA* 4.4, Aristotle begins an account of various different kinds of monstrosities and deformities. One might think it pretty straightforward, from previous experience with his writings, that he would have very little to offer here in terms of scientific explanation. Monsters do not occur always or for the most part, but infrequently. Surely by being contrary to nature, they are the result of luck or chance (*Phys.* 2.5, 196b10–17, *Meta.* 5.30, 1025a14–16), which are accidental rather than *per se* causes and so not properly explanatory (*Phys.* 2.5, 197a33). Furthermore, rare events are associated with incidental ones and it is not possible to know about incidental matters (*Phys.* 2.5, 197a19–20, *Meta.* 6.2, 1027a19–23, and *APo* 1.30, 87b19–27). The detailed accounts that we find in *GA* 4.4 show Aristotle attempting to explain why monstrous and deformed animals occur, and so seem to count as explanations in the realm of his science, broadly construed. Is there any way to provide a properly scientific explanation of why sometimes monsters come to be? In this essay, I argue that scientific knowledge of monsters is not possible. As with other chance events, two separate causal stories can be used to further explain what happens,[1] but there cannot be any *per se* causes of monsters. Given this, it is then perhaps surprising that Aristotle spends so long on monsters and related deformities. In what follows I will set out the account in some detail and show that, although there can be no scientific demonstration or knowledge of monsters, there are many recognizable elements of scientific explanation in *GA* 4.4. What happens in cases of monsters and deformities occurs in the process of generation, and there is much that we can know scientifically about this process – working from the animal's essential attributes outward to factors that influence these processes. In particular,

[1] As in the case of a man who meets a creditor by chance in the market (*Phys.* 2.5, 197a16–24).

we find Aristotle looking for and investigating whether "for the most part" (FMP) correlations are causally relevant and linked to essence. Even though the birth of monsters is scientifically inexplicable, the features of animals that tend to produce them can be known about. I will explain how it is that producing more than one offspring per birth is a knowable feature of a kind. This feature, along with others, can then give us a much better grasp on what happens in particular instances of monstrous birth.

I will begin by setting out some of the explanatory machinery that Aristotle must have in mind as he approaches the challenge of giving an account of monsters and individual deformed animals in *GA* 4.4. I will then give an overview of what he says about three key types of problem birth – monsters which are two animals grown together, deformities that involve redundancy of parts, and deformities that involve deficiency of parts. I will finally more closely consider three aspects of this account that can help us to assess its scientific merits: Aristotle's use of regularity, his search for relevant differentia, and the resultant findings about the feature of bringing to birth many offspring at once. I conclude by considering whether there is any way to include monstrous birth in Aristotle's science through the idea that "the way is prepared by nature" (*GA* 4.4, 770b4) for them.

Scientific Explanations of Natural Phenomena

The *APo* provides a paradigmatic account of scientific knowledge (ἐπιστήμη) in terms of necessary demonstrations where the middle term picks out the cause of the necessary relationship between a subject and its *per se* attribute (e.g., *APo* 1.13, 2.1–2).[2] This model is refined in order to accommodate biological knowledge.[3] Aristotle also distinguishes three types of events and explanations of them. First, there is absolute necessity, which holds always – encompassing eternal cyclic processes (*GC* 2.11, 337b30–338a18). Next, there are those things that hold only FMP, including the objects of biological science. Finally, there are incidental events, which we think of as the result of chance or luck.

> There is no understanding through demonstration of what holds by chance. For what holds by chance is neither necessary nor for the most part, but what comes about apart from these; and demonstration is for one or other of these. For every deduction is either through necessary or through for the most part propositions; and if the propositions are necessary, the conclusion

[2] See also Lennox 2001b: 128–31. [3] Lennox 2001b: 128–31; Leunissen 2010: 43–5.

> is necessary too; and if for the most part, the conclusion too is such. Hence, if what happens by chance is neither for the most part nor necessary, there will not be demonstration of it. (*APo* 1.30, 87b19–26. trans. after ROT; cf. *APo* 2.13, 96a7–11)

Events occurring due to chance, such as monsters, are immune to scientific knowledge but natural events that happen FMP can be demonstrated and known.[4] One important resource for understanding the flexible and complex model of scientific explanation is a distinction between causal and modal uses of necessity (*APo* 2.12). Causal necessity in sublunary natural objects is conditional necessity.[5] In sublunary phenomena, the relationship between the subject and its attribute is not absolute but conditional and the middle term is an expression of teleological causation (*PA* 1.1, 640a1–9, 33–b3). Modal necessity is the idea that the relationship between cause and effect is not absolute but only occurs FMP – which is particularly useful in the investigations in biology where internal and external material factors can impede goal-directed natural processes in particular instances (*PA* 1.1, 639b29–31, *GA* 4.8, 777a18–21).

Finally, demonstrations require starting points in the form of axioms and posits (*APo* 1.2), which are further subdivided into suppositions and definitions. Definitions will state the essence of kinds (*APo* 1.10). The search for definitions is evident in many parts of the zoology,[6] as is the search for essential characteristics of animals. Aristotle looks for features of animals that occur regularly together so that in a second stage of analysis, he can attempt to discover if there is any necessary (in the modal sense) link between the features.[7] Devin Henry (2015: 172–74) helpfully classifies three types of FMP propositions found in the biological works. Category A1: *ceteris paribus* laws; Category A2: the more and the less; and Category B: correlation without causation. A1 expresses a defeasibility condition; for example, if females FMP do not menstruate when lactating, this means they do not unless material factors intervene or impede what normally occurs (*GA* 4.4, 777a3–21). A2 expresses the fact that some natural features are plastic but tend to cluster in one area – so that although dogs FMP begin to lactate five days after birth, it is natural for them to do

[4] See also Henry 2015: 171.

[5] The distinction was recently brought into focus by Kupreeva 2010 and refined by Leunissen 2010a: 46–50.

[6] In particular in the *HA* and *PA* but also in the *GA*. See his attempts to define male and female (*GA* 1.21, 729b8–19, 766a31–4).

[7] See Henry and Nielsen 2015: 6, and Henry 2011: 211–12.

this between four and seven days after (*HA* 6.20, 574b6–7). B-type cases are ones in which sometimes FMP propositions are merely "observable regularities" with no underlying causal basis, such as the correlation of having many toes with producing many offspring (*GA* 4.4, 771b2–8). The most straightforward type is A1, which can be used in scientific demonstrations, since propositions of this type can be easily reformulated so as to hold in every case (e.g., "all normal pregnant female animals necessarily lactate if nothing impedes").[8] A2 cases are more difficult since just because something happens FMP in this way does not mean that cases that fall outside that but within a normal range are not also natural.[9] It is clear, though, that B-type propositions will be useless in our search for scientific knowledge.

We can see, then, that biology contains modes of explanation originating in an *Analytics*-style structure, inevitably reformulated for use in the sublunary realm where entities are changeable and contingent. The proper knowledge of them requires that we find the links between the *per se* attributes of entities in order to come up with informative demonstrations.

Aristotle on Monsters and Deformed Kinds in *GA* 4.4

In *GA* 4.4, Aristotle concentrates on (i) monsters which are two animals grown together, like conjoined twins; (ii) deformities that involve redundancy of parts – one animal with parts of two or more; and (iii) deformities that involve deficiency of parts. The section is well crafted to discuss these three types of monstrous birth, linking them through a common set of causes. It ends with the related issue of superfetation, which is also the result of similar causes.

Aristotle seeks a demonstration using the same cause for various phenomena surrounding the generative processes, in particular instances including sexual differentiation, hereditary resemblances, and monstrosities (*GA* 4.3, 769b3–9). It is clear from the earlier part of *GA* 4 that one of the main causes in the case of sexual differentiations and heredity is the state of the mixture of the male and female generative residues. In *GA* 4.4, a broad statement of the importance of συμμετρία, due proportion (767a14–35), is followed by an intricate exposition of what goes on in this mixture. An individual animal comes to be male or female and to resemble its relatives in certain parts and features due to the δυνάμεις and the κινήσεις

8 Henry 2015: 183–4.

9 *Ibid*, 184 attempts to reformulate FMP propositions of type A2 through another essential feature.

derived from these battling it out (*GA* 4.3, 767b21–768b15). And when everyone loses (including ancestoral κινήσεις reached by relapse), nothing is left to shape the animal appropriately and so it is a monster (*GA* 768b27–9). After a confusion of shaping κινήσεις all that is left is just an animal, not recognizable as a particular kind (*GA* 4.3, 769b13).

Initially, in *GA* 4.4, Aristotle rejects the idea that deformity is due to the exclusively male semen (76931–770a5), opting instead for the view that the problems occur "in the fetations (κυήμασι) as they become established" (770a8). Fetations are a combination of male and female contributions to generation; the same causes determine how this mixture will turn out (*GA* 4.3, 769b4), an idea further elaborated as the proportional relationship between the two residues with appropriate levels of complementary δυνάμεις (*GA* 4.4, 772a2–30).[10] The fetation is where the problem occurs, but in *GA* 4.4 the explanations that Aristotle reaches for do not only involve the intricacies of that mixture but instead broaden out to consider the type of animal in which these problems occur and the generative processes in them, linking these to other aspects of their nature.

In seeking the cause for these three types of problem birth, Aristotle begins to focus on one particular feature of certain animals – the propensity to produce more than one offspring per birth or polyparity.[11] The most polyparous of all are the many-toed (πολυσχιδής) kind.

> [Monstrosities occur] more frequently in goats and sheep, because they are polyparous; and still more frequently in the many-toed, because these animals are polyparous and the offspring is not perfected when born (e.g., the dog) – most of these animals' young, of course, are born blind. The cause why this occurs and the cause why they are prolific must be stated later. But the way to the production of monstrosities has been already prepared by nature (προωδοποίηται τῇ φύσει) by the fact that they generate offspring which, owing to its imperfect state, is unlike its parents. (*GA* 4.4, 770a36–b5, trans. after Peck)

In this most unusual phrase, "the way has been prepared by nature," Aristotle links the animal's nature, its characteristic way of life, to the more frequent occurrence of monstrous birth in it. If, as Aristotle insists elsewhere, monstrous birth is contrary to nature, how can it be that nature has prepared for it? Aristotle continues as follows:

10 This passage more specifically explains why many offspring can be produced at once.

11 Aristotle appears to distinguish the production of many offspring per birth (πολυτοκία) from the production of many offspring over a lifetime (πολυγονία, which we might call "profilicacy"), although the terms are somewhat fluid in the *HA*. Birds have both features (*GA* 3.1); they are polyparous because they produce eggs with two yolks (*GA* 4.4, 770a14–16).

> A monstrosity belongs to the class contrary to nature not in its entirety but only to nature in the generality of cases (τὴν ὡς ἐπὶ τὸ πολύ). So far as concerns the nature which is always, and is by necessity, nothing occurs contrary to that; no, unnatural occurrences are found only among those things which occur as they do in the generality of cases, but which may occur otherwise. (*GA* 4.4, 770b9–13, trans. after Peck)

This statement makes a familiar point, reminding us of the contrast between absolute and modal necessity in the sublunary realm, where teleological ends can be impeded and connections are defeasible. There is nothing here yet to explain why nature itself is involved in preparing for a monstrous birth. It is only through the next statement that we are further enlightened.

> Even in those instances of the phenomena we are considering, what occurs is contrary to this particular order, certainly, but it never happens in a merely random fashion (μὴ τύχοντως); and therefore it seems less of a monstrosity (ἧττον εἶναι δοκεῖ τέρας) because even that which is contrary to nature is, in a way, in accordance with nature, that is, whenever the formal nature has not gained control over the material nature.[12] Hence, people do not call things of this sort monstrosities any more than they do in the other cases where something occurs habitually (εἴωθε) [...]. Thus, there is a certain sort of vine – "smoky" is the name some people give it; and if it bears black grapes they do not reckon it as a monstrosity, because it often and habitually does this. (*GA* 4.4, 770b13–22, trans. after Peck)

Because it is natural for the many-toed to produce offspring that are incomplete, this means that in some cases monsters are not "by chance." In the first line, Aristotle refers to "the phenomena we are considering," and at the end of this passage back to the monstrous conjoined offspring (*GA* 4.4, 770b25), and so is indicating that these monsters are part of the natural order in some way. He does not simply use the idea that natural processes are defeasible and inexplicably go wrong every once in a while. This would, in fact, allow him to give a statistical account of why monsters occur more in the polyparous. Say that a monster occurs once in every 5000 births – if the monoparous elephant ends up producing only one offspring, then it is much less likely to have a monster (a one in 5000 chance) than an animal which produces 100 offspring in its lifetime (a one in fifty chance).[13] But Aristotle does not say this. Instead, he is interested in the characteristics of this sort of animal and how it reproduces.

[12] *Phys.* 2.8, 199a31: nature is twofold: matter and form.

[13] This explanation would, in fact, be more appropriately applied to all prolific animals, whether polyparous or not.

The reasons for increased frequency of monstrous birth begin with polyparity; when compared with prolificacy on its own, it presents an extra risk factor as the many internal offspring can hamper each other's completion, confusing the usual κινήσεις in development (*GA* 4.4, 770b28).[14] The many-toed also add another challenge, that is, their relative lack of vital heat. This means that the female parent is unable to complete the offspring at the point of birth (*GA* 4.4, 770b4–5), making their growth onto each other even more likely (*GA* 4.4, 774b10–17).[15]

Aristotle next expands the discussion to include two other closely related birth defects – the production of redundant parts and a deficiency in parts, both of which are also in some sense caused by polyparity (*GA* 4.4, 770b28, 771a14). Rather than a focus on what happens in all instances of generation, where the state of the mixture can lead to a confusion or weakening of the formative powers, the emphasis here is on the conditions of conception and gestation in particular types of animal. In order to tie all these happenings together Aristotle further explores the reasons for polyparity, seeking a firmer grounding in the animals' natures.

Those that suffer from the most deformities are those that have many toes and this was explicitly linked to polyparity. However, foot type cannot explain polyparity (*GA* 4.4, 771b2–8) as the correlation is neither necessary nor sufficient. Types other than those with many toes are polyparous (e.g., the pig; *GA* 4.6, 774b24–26) and some animals with many toes are monoparous (e.g., the elephant; *GA* 4.4, 771b8–9). Furthermore, the frequent occurrence of the two together is a mere "observed regularity" without any causal basis.[16] Instead, the number of offspring per birth is determined by two features that are fixed in the nature of animals – their size and their relative degree of vital heat.[17]

Larger size is the reason why (διά) an animal is not polyparous – since nourishment is used up on bodily growth and so there is less residue to become the seminal residue (*GA* 4.4, 771a24–25). Smaller animals, then, produce more semen and therefore can make up more offspring at a go.

[14] Aristotle does not explain the absence of the sorts of membranes between offspring that he says at *GA* 2.7, 746a12–19, prevents such conjoining.

[15] This further clarification happens in *GA* 4.6. The heat of the female animal's body is necessary to complete the offspring, suggesting that the male semen is not alone responsible for all differentiation of the new animal, a point I have argued elsewhere (see Connell 2016).

[16] See also Henry 2015: 174.

[17] Aristotle proclaims at several points that we ought not to be amazed (θαυμάζειν, *GA* 4.4, 771a19–27) by the correlation of parity and body size. On this correlation, see also *GA* 1.18, 725a28–34. On body size as an essential feature of a kind, see *PA* 4.6, 683a18–19; *PA* 4.9, 685b12; *GA* 2.6, 745a5–9; *DA* 2.4, 416a15–18, *IA* 8, 708a9–20; and *Pol.* 5.9, 1309b18–35.

But since Aristotle rejects the idea that regions of the uterus draw in semen (*GA* 4.4, 771b27–33),[18] he must explain why more generative residue doesn't just produce one large embryo (as more milk and fig-juice produces a larger block of cheese, *GA* 4.4, 771b25).

> When animals are being perfected, there is a certain size for each, a limit of bigger and smaller; none will be formed bigger or smaller than these sizes [. . .]. In precisely the same way, the seminal matter [i.e. the female contribution] out of which [the embryo] is formed is not unlimited in either direction – the amount of it can be neither bigger nor smaller than certain limits; the embryo cannot be formed out of any casual amount of it. Thus, in the case of those animals which (on account of the causes stated) discharge more residue than is requisite for the principle of a single animal, it is not possible that the entirety of this should be used to form one embryo; on the contrary, as many are formed as is determined by the sizes proper to those animals. (*GA* 4.4, 771b33–772a8, trans. after Peck)

Aristotle further explains, on the basis of the essential size limits of a given kind, why when larger animals have a good amount of generative residue, this is not divided up to make many offspring at once:

> [In] those animals which are large and produce one offspring only, a large amount of residue does not give rise to a large number of offspring, for the same holds good: here too, the amount of the material and of that which works upon it are definite. (*GA* 4.4, 772a30–33, trans. after Peck)

We now believe that each embryo is roughly the same size – i.e., one cell/ovum. Aristotle, in contrast, must hold that even brand-new embryos differ in size. Although counterintuitive to us, there is nothing theoretically implausible about such a stance, which explains why a large amount of generative residue only results in many offspring if the parents are small-bodied.

The same reasons lie behind deformities that involve redundant parts and those that involve deficient ones, i.e., the animal's size and the amount of potent generative residue it produces (both male and female). In the case of redundant parts, the male and female have produced enough generative residue to constitute more than one offspring, but not enough to make two: the materials then ooze out to form more parts than are required (*GA* 4.4, 772b13–26). About deficiency of parts, Aristotle has very little to say, only that they are like spontaneous abortions – presumably, in these cases there is not quite enough generative residue to complete the

[18] The Pseudo-Aristotle *Problemata* 10.41 considers this theory, which appears in the Hippocratic treatise *On Nature of the Child* [*Nat. Puer.*] 31 (Potter 2012: 540), to be viable.

instrumental parts. The difference between the main cases Aristotle focuses on in *GA* 4.4 are then summarized: monstrosities are the growing together of two animals, whereas deformities involving redundancy are one animal with too many parts (*GA* 4.4, 773a8–13).

Explanation and Knowledge in *GA* 4.4

This section will more closely consider three related issues in Aristotle's explanations of monsters and deformities. I will begin by clarifying the role of frequent as opposed to rare occurrences in Aristotle's scientific explanations. Next, I will look at Aristotle's search for relevant features of kinds based on initial clues that relative frequency provides. Finally, I will focus on the feature of the number of offspring produced per birth to see how it might fit into Aristotelian science.

Within *GA* 4, Aristotle is repeatedly looking for and finding causes.[19] The framework of these causes is complex. As mentioned, it seems that one set of explanations focuses on the state of the mixture where Aristotle provides numerous posits (ὑποθήσεις) such as the presence of current (ἐνέργεια) and possible (δύναμις) potential κινήσεις,[20] the idea that being mastered results in changing into the opposites, and the workings of relapse (*GA* 4.3, 768b5–10).[21] In *GA* 4.4, there is an explicit drive to discover a demonstration of the causes of monstrosity (769b11). Although the state of the mixture is certainly still part of his explanation of monstrous birth, Aristotle focuses now more on animal kinds and their characteristics.

In order to advance his knowledge of nature, Aristotle believes that he must find the *per se* attributes that attach to a natural subject through teleological reasoning. For example, in order for a person to be healthy, certain states of affairs must hold; these states of affairs are conditionally necessary for her health (*PA* 1.1, 639b30–40a1). This holds FMP as it is a sublunary phenomenon and so defeasible. Finding items that hold together FMP is an important first step in attempting to discover more fundamental connections and explanatory *per se* attributes of a kind. But something's happening frequently or rarely is no guarantee of such

[19] E.g., *GA* 4.1, 766a18; 4.1, 766b26; 4.2, 767a14–5; 4.3, 768b16; 4.3, 769a1; 4.3, 769b3; and 4.3, 769b11.

[20] For an explanation of my preferred translation, see Connell 2016: 311–2, 316, 319, and 322.

[21] The section on sexual differentiation also lists a series of premises (ὑποκειμένων, *GA* 4.1, 766a16), which must be accepted before the reader can understand the causes of male and female coming to be.

discoveries. There is no necessary connection between frequent occurrences and natural teleological ones in Aristotle's science.[22] Some events that happen infrequently may still be natural, such as the birth of a human child. Such occurrences fit with a "conditional frequency" model – under given conditions, things will happen FMP in a certain manner.[23] Although it is rare, a woman who has a child will FMP have a human child. Even when we have narrowed down such FMP propositions, there are still cases where the relevant causal connections are missing. I propose that Aristotle is well aware that correlations between events or features of a kind are not necessary or sufficient to count as properly explanatory but could be an indication or sign of these and so are worth investigated further. He can then get to work testing each until he can fix on the best and most scientific explanation and one that could ultimately feature in demonstrations. Those FMP correlations that turn out to be more promising will require careful specification, in terms of both the plasticity of the feature and those factors which can impede or prevent it. Returning, then, to Henry's three Categories of FMP propositions, we can apply these to Aristotle's investigations in *GA* 4.4 to see how far he gets.

FMP correlations will initially guide Aristotle in his attempts to bring to light which features of a kind are essential to it and so appropriate for defining kinds and providing scientific explanations. In *GA* 4.4, type of feet might at first have looked promising. Those that are prone to produce deformed offspring tend to be polyparous and almost all of the many-toed are polyparous. Having many toes[24] is correlated with other differences in the biological corpus, such as the shape of the spleen (*PA* 3.12, 674a1) and the absence of horns and hucklebones (*PA* 3.12, 674a1, *HA* 2.1, 499b23, and *PA* 4.10, 690a24). However, a focus on feet is a false start: although many toes and polyparity tend to occur with great frequency together, they are not causally related. Furthermore, no proper explanation can be attained by leaving the matter at the fact of polyparity. Having more offspring at a go only occurs because more generative residues are produced and this abundance of residues is not there for the sake of polyparity. It is, rather, a material consequence of the size and body heat of the

[22] See Judson 1991: 86, 97. [23] *Ibid.* 89.

[24] This is a division within the blooded and viviparous kind. Cf. *HA* 2.1, 499b6–11: "Of blooded and viviparous quadrupeds some have the foot cloven into many parts, as is the case with humans [. . .] (for some animals are many-toed, as the lion, the dog, and the leopard); others have feet cloven in two, and instead of nails have hooves, as the sheep, the goat, the deer, and the hippopotamus; other are uncloven, such as the solid-hoofed animals, the horse and the mule" (trans. after Peck). See also *HA* 2.10, 502b34; *PA* 4.10, 690a5–7.

kind which results in the production of plentiful and potent residues. The residues themselves are telic but the amount is contingent on many other factors, such as external weather conditions.

So neither foot type nor polyparity are of much use on their own in this search for knowledge. In a later work in Aristotle's school, the *Problems,* the author starts immediately with body size as the essential factor to consider with respect to the issue, asking "why is it that quadrupeds of a small size most often give birth to monstrosities, whereas man and the larger quadrupeds, such as horses and asses, do so less often?"[25] The tentative reply features both sets of causes considered in *GA* 4.4, polyparity and the confusion of seminal fluids.[26] Although there is an overlap here in the content of *GA* 4.4 and the *Problems,* there are also significant differences. In the *GA*, Aristotle is not merely speculating about whether there is a connection between certain deformities and polyparity – we find instead a more systematic attempt to explain this connection. Aristotle's focus on body size and relative degree of vital heat fixes the explanations to the animal's essence. The combination of these two factors gives the reasons why the animal has the number of offspring it does per birth. Smaller size and a high level of vital heat means that a great deal of potent generative residue goes to make up many smallish embryos and the heat ensures that they grow and develop properly.

The fact of polyparity is not an essential feature and so it does not appear to be explanatory. It is not conditionally necessary; it does not happen for the sake of anything but rather comes about because body size and vital heart interact in a particular manner in the process of generation. As this, it is the result of two such crucial features; however, it occurs with enough regularity to indicate naturalness and a loose connection to the kind in question.

We might at this point profitably consider how the correlations Aristotle has so far discovered fare in terms of the requirements for adequate demonstration proposed by Devin Henry.[27] Size and degree of vital heat are most promising as features of kinds. An elephant is FMP large (unless something impedes). Meanwhile, a human being FMP has a high degree of vital heat (unless something impedes). Bodily size and degree of vital heat are clearly excellent features for Aristotle to concentrate on in his biological explanations. As for number of offspring per birth, the answer is more complicated. Although it is not essential to a kind, it may still be feasible to use it in scientific explanations, since it is fixed within certain limits.

[25] *Problems* 10.61, 898a9. [26] *Problems* 10.61, 898a10–15. [27] Henry 2015.

This is because it can be tied causally to size and degree of vital heat. Thus, size is FMP correlated to the number of offspring per birth. The connection is defeasible due to fluctuations in the amount of generative residue produced but this in turn can be fixed with reference to the degree of vital heat FMP possessed by the animal kind. This means that the number of offspring per birth will often be fixed for a kind. So, for example, an elephant will FMP produce enough generative residue to generate one elephantine embryo. For some kinds, however, this is not the case. For human beings in particular, it is more difficult to fix how many offspring they will produce per birth, as the FMP propositions associated with twinning are more like Henry's A2. This may help to explain why Aristotle seems not to have made up his mind about whether human twinning is natural or not.

Sometimes Aristotle says that for humans having twins counts as a monstrous birth and that having one offspring is most natural to them (*GA* 4.4, 772a36–772b1). He also says that human dualize (ἐπαμφοτερίζει), sometimes producing one and sometimes more than one offspring per birth (*GA* 4.4, 772b1–3);[28] it would seem that both are natural to humans. It might even be the case that "humans are by nature polyparous" (*GA* 4.5, 773b23–24). Certainly, human twinning is rare but this need not make it unnatural. Henry's A2 FMP phenomena do not necessarily pick out the only natural happenings within a more or less continuum. It is natural for a dog to lactate seven days after birth, even if FMP dogs lactate after five. So also it may well be natural for a human to have twins even if FMP humans do not do so. After all, they have a good level of vital heat and so are full of residues (*GA* 4.4, 772a36) and have a roomy enough uterus (*GA* 4.5, 773b24), all features which are perfectly natural to them.[29] On the other hand, Aristotle might choose to limit the range of what counts as natural in this case, unlike the dog case, specifying that one offspring is what is natural (*GA* 4.4, 772b3). His decision will no doubt be based on other factors he takes to be relevant. He might, for example, consider the places and sorts of human beings that more regularly produce twins and decide that they are unnatural and freakish.[30]

[28] See also mules and deer (*HA* 6.22, 576a1; *HA* 6.29, 578b15).

[29] Another feature of human beings also aids multiple pregnancy, that is, their variable gestational period (*GA* 4.4, 772b8–10). An animal's gestational period is said to be proper to it (οἰκεῖος) indicating that it is likely to be a part of its essence (*GA* 4.3, 769b25).

[30] Aristotle believes that people living in places other than Greece are somehow defective (*Pol.* 7.7, 1327b29–1328a14). Twins occur most in Egypt (*GA* 4.4, 770a35; *HA* 9(7).4, 584b31–32).

Returning, then, to the number of offspring per birth as a feature of a given animal type – it can be fixed by nature, at least for some animals. The fact that it is natural can then make better sense of two anatomical facts. Polyparous animals will need larger uteri and more numerous nipples, and nature does appear to supply them with these.[31] The size of the uterus is fixed by nature to suit the size and number of the offspring. One could argue that this is not the case, given that it initially comes to be in order to contain the female generative residue (*GA* 4.1, 766b23–25), with no mention of gestation. Indeed, some uteri would seem to always be a bit too big, such as those in animals that superfetate. These animals have left over space in their uteri even after there is an embryo in it (*GA* 4.5, 773b24). Even if the size of the uterus may not necessarily be for the sake of polyparity, the number of nipples an animal has is (*PA* 4.10, 688a31–b25), as Aristotle makes clear.

> Polyparous animals have their mammae upon the abdomen. This is because they have numerous young to feed, and so they need numerous mammae. (*PA* 4.10, 688b15–17)

We can now establish the relevant sense in which polyparity is natural to a kind and subject to scientific understanding. It is not directly teleological but connected through essential features of the kind. It is more variable and plastic than other features, fluctuating depending on a number of factors such as time of life and locality. It comes along with essential features, in particular body size and amount of vital heat, which then result in abundant and potent generative residues. This, again combined with small body size, means that many embryos can make a start in the uterus all at once. One might think of it as a feature that follows the pattern of secondary teleology, where nature uses material conditions to further its aims.[32] This seems to be plausible, since one of the goals of nature will be for each individual to reproduce itself as many times as is possible.[33] However, one might be reluctant to characterize this feature as so closely tied to teleology since it leads to detrimental results, i.e., an increased risk of severe deformities.

Can any monstrous birth, if regular and explicable in terms of animal kind and process of generation, fit into Aristotle's natural science? The best chance for this is those conjoined offspring that most frequently occur in

[31] Even though a small uterus will be able to contain many undeveloped embryos there would seem to have to be some natural mechanism to prevent more than the viable number forming there, some "foresight" about how large the embryos will eventually get.

[32] On "secondary teleology," see Leunissen 2010: chapter 4.

[33] Connell 2016: 240–3.

the many-toed kind. In these cases "the way is prepared by nature" which means those monsters are "not by chance" (*GA* 4.4, 770b15). Aristotle suggests, then, that the conjoining of twins in the many-toed kind is "not monstrous," just as black grapes on the so-called smoky vine aren't (*GA* 4.4, 770b20).[34] If they are not really monstrous, then it may be that we can find a proper scientific explanation for them. The way he seems to try to do this is through the idea that the monstrosity of the offspring is somehow a part of the nature of the kind – because they produce monstrous offspring almost all the time. In the many-toed, their offspring are *never* complete and so are deformed already before the growing together even happens. "The way is prepared by nature" for monstrous births in this kind because due to middling vital heating capacities, they never complete their offspring, which are born "deformed," i.e., unarticulated (*GA* 4.6, 774b10–17).[35]

If this is how Aristotle wants to include deformity within natural explanations, then he is going to find this a tricky position to maintain. Indeed, it does not fit well with his overall teleology, which makes it very difficult to see how enforcing systematic deformity could be a part of nature. Rather, nature always attempts to achieve what is best and in kinds like the many-toed, its failure, even though frequent and in a sense explicable, is not natural or part of the natural order. A good way to make sense of Aristotle's deformed kinds has been put forward by Charlotte Witt.[36] Looking particularly at aquatic quadrupeds, lobsters, moles, and females,[37] Witt concludes that to be deformed for Aristotle is the impeded development of a part in embryogenesis. This results in an animal with a part that is (i) incomplete and therefore not functional or (ii) not used for its proper function.[38] So the mole has eyes that are incomplete and nonfunctional, while the seal and lobster use parts for roles they are not designed for (the seal swimming with a walking instrument and the lobster walking with a grasping instrument).[39]

[34] Nielsen 2008: 394 provides an interesting interpretation of this passage. For her, Aristotle is arguing against the common view that black grapes are not monstrous; instead he thinks they are, just like female births are.

[35] See also *HA* 6.33, 580a4–5; *GA* 2.6, 742a1.

[36] Witt 2012.

[37] Allan Gotthelf (1985: 39–41) gives a different account of deformity of kinds, where the kind is deformed relative to the wider kind to which it belongs.

[38] The Greek terms used for deformity are often used of people without limbs, who then might have to use other limbs in order to complete daily tasks. See Witt 2012: 90–1.

[39] I disagree with Witt that the female is characterized by Aristotle in the same way, as lacking a functional part. The parts that are unfinished, defective, or used for other purposes in the seal, lobster, and mole are instrumental parts. For Aristotle, male and female do not fundamentally differ

This account avoids the difficulty of having a teleologically focused form that is systematically producing defective animals, instead positing that the errors occur in the process of embryological development in each instance. This makes better sense of the term "ὥσπερ" used of the deformity of such animals – they are like animals that are deformed in the process of generation in other kinds – like the blind infant, whose eyes never developed properly. A potential difficulty for this theory is that quite a few nondeformed animal kinds are said to use their instrumental parts for dual functions. So the elephant uses it "nostril" as a hand (*PA* 2.16, 659a1), and this is because, unlike other quadrupeds, who use their front feet as hands, the elephant cannot do so due to its weight. As Aristotle says, nature in this case is following her wonted plan of using one and the same part for several purposes, for in the many-toed quadrupeds the fore feet are intended not merely to support the weight of the body, but to serve as hands (*PA* 2.16, 659a21–25; cf. *PA* 4.10, 687b1 and 688a2–5).[40] Although there may be challenges with applying case (ii), case (i)[41] seems very apt and applies well in the case of the many-toed kind. Rather than the production of unarticulated animals being a part of their form or nature, it is the case that their form systematically fails to establish itself in order to complete the offspring at the point of birth. And there is nothing natural about this. Thus, there isn't any way in which the deformities are natural or subject to any natural teleological explanation based on *per se* causes. Instead, the explanation will be in terms of material factors, divorced from essence; it will concern the state of the mixture of generative residues. We can conclude, therefore, that though conjoined offspring occur more often in the many-toed, this does not make them less monstrous but only makes it seem that they are less so. It is perhaps this false appearance of naturalness that leads Aristotle to remark that they are "not by chance."

with respect to their instrumental generative parts but due to the heating action of their heart (*GA* 4.1, 766a35–b5). This is the part that forms before the rest and directs the development of the other parts (*GA* 2.6, 742a16–b18; cf. *GA* 4.4, 773a5–6). Any serious infirmity in this part would mean that the animal could not develop at all. For further discussion, see Connell 2016: chapter 8.

40 See also the dual function of female mammaries (*PA* 4.10, 688a16ff; and b24–25) and the octopuses use of an arm as a penis (*HA* 4.1, 524a8–10).

41 This type of deformity, a deficiency of the parts (770b30; 772b37–773a2), is the one that Aristotle has the least to say about in particular instances in *GA* 4.4. This is probably because these embryos are much less likely to survive, particularly if the part that is deficient is a controlling part. They are often spontaneously aborted (*GA* 4.4, 773a1–2).

Conclusion

In investigating the causes of monstrous births, Aristotle seeks some kind of demonstration (*GA* 4.3, 769b4). Moving away from focusing on the mechanisms at play in the mixture of generative residues, he broadens his analysis to consider the types of animals involved and the features that are essential to them and explanatory of their essence. In doing so, he might indeed be getting closer to demonstrative scientific knowledge. By seeking to further investigate observed regularities, we can see Aristotle trying to find out which will move us toward scientific knowledge and which will not. The explanatory framework he settles on provides a backdrop upon which the scientist can situate the occurrence of monstrous birth. So, although she cannot provide any demonstrations of why it happens, since it is anomalous, incidental, and unnatural, she can see when it is more likely to happen based on her knowledge of the various kinds of animals and their essential features. By tying it to essential features, Aristotle can then give an account of a feature that has strong connections to the relevant natural processes. This feature is polyparity, a feature that most often creates the possibility for the sorts of monsters and deformities he is most interested in. Polyparity is the cause of these happenings but since it is not essential, it cannot be linked in the proper manner to the animal kinds in which it occurs, and probably cannot feature as a premise in a scientific demonstration. However, since Aristotle regards it as a pretty stable feature of a kind, piggy-backing on essential features and the regularity of the workings of internal and external material necessity, it can count as "natural." It is tied closely enough to the animal's nature for other anatomical features to be in place such as a given number of nipples and a spacious uterus.

One could see polyparity as in some ways similar to other features Aristotle discusses in *GA* 4 and 5, specifically inherited features and accidental ones. Inherited features follow particular patterns determined by the κινήσεις in the parents' generative residues. When you end up looking like your mother, the κινήσεις won't change into any chance thing (οὐκ εἰς τὸ τυχόν, *GA* 4.3, 768a3) but only into the opposite of your father. Although they are not present for the sake of anything, we can think of them as "*per se* results of certain κινήσεις, which are said to be drawn from potentials that the generator has nonaccidentally, *qua* generator."[42] We might also here bring to mind the accidental features from

[42] Gelber 2010: 191. See also Connell 2016: 320–1.

GA 5 (e.g., eye color). Owls have to have eyes constructed in a particular manner from particular material (conditional necessity) in order to see acutely. The color that results from those materials conditions is not part of the essence (*GA* 5.1, 778a33, and 779a30–31) but will invariably appear. The pattern of consistent eye color following from an animal's essence is similar to the consistency with which polyparity follows from essential features.[43]

In the case of number of offspring per birth, it would seem that natural causes underlie and make it so that although these are not *per se* causes (not intended directly), nature (like deliberation) can somehow predict that these concurrences will happen and in a way intend them.[44] What nature, however, can never intend is any increased likelihood of monsters and deformity. So although they can be predicted, monsters are the result of pure chance. There is a natural grounding for them but we cannot have any proper knowledge of monstrosity itself other than that it involves the absence of teleology (*GA* 4.10, 778a5–9).[45] And so we are forced to return to the messy details of what occurs in the mixture of male and female generative residues in each particular instance of generation (*GA* 4.3, 768b28–35) in order to seek reasons for them.

[43] In the case of eye color, however, the color itself is nonteleological. Many offspring per birth, on the other hand, can find a connection to the goal of generation.

[44] Allen 2015: "It seems conceivable that there could be cases in which by doing one thing *per se* one could hardly fail to do something else *per accidens*, as a concomitant. Elsewhere, of course, Aristotle is happy to allow that a substance is always and necessarily attended by some of its accidents (*Meta.* 1024a23–4; *APo* 1.7, 75b1) [. . .] to the well informed agent, they come as no surprise, and they are part of what he does intentionally even if not with the intention of bringing them about" (79).

[45] Johnson 2005: 198.

PART IV

Methodology in Aristotle's Generation of Animals

CHAPTER 12

The Search for Principles in Aristotle Posterior Analytics *2 and* Generation of Animals *1*

Robert Bolton

Aristotle's Search for Principles

Aristotle begins his studies in natural science, in *Phys.* 1.1, with the following remarks:

> Knowledge and scientific knowledge in every mode of inquiry where there are principles or causes or elements comes from the knowledge of these things. For we think we know a thing when we know its primary causes and its first principles and we are back to its elements. It is evident then that in the science of nature too the primary task must be to try to settle things concerning the principles. (184a10–16)

Aristotle goes on in *Phys.* 1.1 to offer a sketch of how principles in science do become known. His proposals there are much more fully developed in the *APo*, especially in Book 2. In *APo* 2, however, Aristotle focuses exclusively, or almost exclusively, on just one type of principle, namely the definition or account of the ultimate essence of a kind and on how such a definition comes to be known.[1] This special emphasis reflects, of course, an inheritance from Socrates and Plato, who themselves say but little about any other types of principles.[2] But Aristotle opposes the Platonic thought, or sometime Platonic thought, that knowledge of the ultimate definitions or essences of things is innate knowledge which needs only to be brought to mind by proper dialectical investigation. He is, however, still sensitive to a main point which helps to lead Plato to take that view seriously, namely that to come to have any other genuine knowledge about a kind one must first know what that kind is. For how

Earlier versions of this chapter were presented at Western Ontario, Paris IV-Sorbonne, and UCLA. I am grateful to the members of the audiences there, and especially to the editors of this volume, for very helpful comments and suggestions.

[1] See, e.g., *APo* 2.17, 99a21–23. [2] *Republic* 6.510 c ff. is one of the few exceptions.

else could one know, or come to know, that any discovered facts concern that very kind if one does not already know what that kind is?[3]

To accommodate this Platonic insight, or concern, while avoiding the posit of innate knowledge of ultimate essence, Aristotle introduces a special new type of genuine definition of what a kind is, other than the account of the ultimate essence of a kind, a new type of definition the grasp of which gives one genuine knowledge of *what a kind is* which is sufficient to enable an inquirer to successfully search for and to discover the ultimate essence of that kind, and thus the definitional principle which states what that essence is, without already knowing, innately or otherwise, what it is.[4] This new type of definition is the one known traditionally as the *nominal definition*. It is introduced by Aristotle in *APo* 2.10 as "an account of what something is," which is also "an account of what a name or other naming phrase signifies" (93b29–31). Since the grasp of a nominal definition can play such an important role for Aristotle in facilitating scientific inquiry leading to the discovery of the ultimate essence of a kind, a role comparable in importance to the role played by the posit of innate knowledge of essence, or its variants, in Plato, it is crucial for an understanding of Aristotle's view of the proper scientific method which leads to the discovery of ultimate definitional principles to know exactly what the character and content of nominal definitions is, how they come to be known and how knowledge of a nominal definition does facilitate the discovery of the ultimate essence and definition of a kind. This issue has in recent decades received much, still expanding, attention in the literature.[5] In this investigation I would like first to consider a family of recently most prominent views on these matters and the basis offered for them in passages in *APo* 2, and then to test further the merits of these recent views specifically by reference to Aristotle's actual procedures of inquiry in his pointed search for definitional principles in *GA* 1.

Aristotle's Search for Definitional Principles in *APo* 2

It will be useful to begin this investigation by considering the now widely influential proposal that is developed in what has been called a

[3] See, e.g., *Meno* 79 D ff. For various accounts of Plato's proposal, or proposals, on these matters, see Bolton 1976, 2012 and, now, Fine 2014, with multiple references there.

[4] For this reading of Aristotle's aims in *APo* 2 see Bolton 1976. For the many later variations on this reading, see Bronstein 2016 with extensive references there, and below.

[5] For full discussion of this literature from Bolton 1976 on see especially now Bronstein 2016. The present essay newly offers a considered response to main elements of that subsequent literature.

"three-stage" account of Aristotle's treatment, in *APo* 2, of how proper scientific inquiry leads to the discovery of the ultimate essence and definition of a kind. Speaking generally, on this three-stage approach such a scientific inquiry, according to Aristotle, is completed at the first stage with a grasp, on the part of the inquiring scientist, of an account of what the *name* of the kind in question *signifies*. Inquiry then reaches a second stage with the discovery of the *existence* of the kind signified by the name in question and a third and final stage with the discovery of the *ultimate essence* of the kind. The main features of such a three-stage account are found already in Ross, who, in commenting on a passage in *APo* 2.8, says this:

> The normal order of events [in scientific inquiry, according to Aristotle,] is this: we begin [at Stage 1] by knowing that there is *such a thing as* [for instance, a lunar] eclipse and that this [i.e., eclipse or the name "eclipse"] *means* some sort of loss of light. We first [i.e., next, at Stage 2] ask if there is any *evidence* that the moon suffers eclipse [previously understood as some sort of loss of light] and we find [in a successful case] that there is, viz. the moon's inability to produce a shadow at a time when there are no clouds between us and it. Later [having so discovered its existence] we find that there is an explanation [and, thus, ultimate essence] of lunar eclipse, viz. the earth's coming between the moon and the sun.[6]

More fully articulated versions of such a three-stage account, or others in this same vein, have more recently been treated by many scholars.[7] While these versions differ in certain ways the most relevant elements of their proposals may be initially captured as follows:

> [According to Aristotle], *every case* of scientific inquiry [concerning a kind] involves a first stage where one need not know of the existence of the kind but *must* know an account of what the name [of the kind] signifies [. . .]. At Stage 1 one will have some idea of what it *would be* for there to be [for instance, lunar] eclipses but *will not as yet* have established that there are any [. . .]. The source of the knowledge [of what it would be for there to be, e.g., eclipses] is clear. It is based on [knowledge of] an account of what the name [. . .] signifies [. . .]. On the basis of a Stage 1 account one knows that the kind, *if* it exists, possesses some specified property. At Stage 2 one discovers that the kind [i.e., now some actual kind] does indeed possess *this property*

[6] Ross 1949: 631, my italics and glosses. It is unclear whether Ross understands the initial first stage account to be always an account of what a *name* signifies or, at least sometimes, simply an account of what it would be to be such a thing as, e.g., a lunar eclipse, if it exists. Later writers vary somewhat on this issue.

[7] See, among others, Sorabji 1980; Demoss and Devereux 1988; Barnes 1994; Charles 1990 and 2000; Bayer 1995; Goldin 1996; Bronstein 2012 and 2016, with further references there.

> non-accidentally [. . .] [and, thereby, one comes] to know non-accidentally that [the kind] exists [. . .]. [The] account of what the name signifies is *general* in form: "Thunder" signifies a type of noise in the clouds [. . .]. [This account] fails uniquely to identify the type of noise in question. While it says that thunder is *a* type of noise, it does not state which [type] it is [. . .]. [Nevertheless] to grasp an account [at Stage 1] of what the name signifies is to have sufficient knowledge to go on to establish whether or not the thing exists [at Stage 2] and what it [i.e., its essence and ultimate definition] is [at Stage 3].[8]

Before looking at the texts cited in support of this reading of Aristotle's scientific method in the search for ultimate definitional principles, it will be useful first to attend briefly to some of its implications as history and philosophy of science – and also as semantics, as some recent writers very usefully do. To facilitate this consider the following scenario: when the first human migrants came into the Americas they and, we may imagine, the inquiring naturalists among them, discovered the existence of, *inter alia*, the red-winged blackbird, the jaguar, the tomato, and tobacco. Again, in more recent times inquiring naturalists working in the Amazon jungle have discovered the existence of a multitude of new species of plant and animal. On the three-stage view, they could not have made these discoveries unless, while "not as yet" knowing of the existence of these natural kinds, they nevertheless already had a name, a specific word or phrase that functioned as a name, for each one of these kinds in their vocabulary, and unless they already grasped an account of what that name signified which articulated what it would be to belong to the kind as a natural kind if it exists, an account sufficient then to direct them to the discovery of the existence and, then, the nature and essence of that natural kind.

Both as philosophy of science, and as history of science as well, these features and implications of the three-stage view are extremely implausible. One does not need to have a name for a kind to discover its existence. Aristotle himself knows this well since he very often refers to actual kinds of whose existence he is aware which are, as he says, "nameless" (ἀνώνυμον, e.g., at *APo* 2.5, 74a8, 21).[9] These kinds, he supposes, have no name, either a one-word name or a phrase that functions as a name, such as, say, "red-winged blackbird," to designate them. That does not, for Aristotle, preclude coming to genuine knowledge of their existence. As a further instance, consider now Aristotle's example from *APo* 2.2 of *night* as

[8] Charles 2000: 30, 39, 35, 44–5, my italics. This influential proposal well captures important core features of the type of view in question. Variations will be noted later in this chapter.

[9] See Bonitz 1870: 69b2ff. for numerous references.

a kind of thing of whose existence one might come to know (90a5). On the three-stage view, to come to know of the existence of night people, including inquiring scientists, were required *previously* to have a name for night, and also, in virtue of that, previously to have a conception of what it would be for there to be night to use in the discovery of the existence of night, before knowing of and independently of knowing of its existence. Again, needless to say, this is extremely implausible. That is not the way people, inquiring scientists included, came to know of the existence of night, or to know in a way sufficient for investigating and discovering its ultimate cause and nature what it would be and is for there to be night. The same can be said concerning how people came to know of the existence of the moon, or of what it would be, and is, for it to exist – another example, from *APo* 2.2, 90a5. Aristotle no doubt believes that *in* or *by virtue of* knowing of the existence of some at first nameless kind such as night or the moon an inquiring scientist can provide some sort of useful description of this kind as a basis for further inquiry onto its nature. But to hold that one *must* have had this description in hand *before* discovering the existence of the kind rather than having come to have it in or by virtue of discovering through experience the existence of the kind would be absurd. Aristotle himself could hardly have known what it would be to exist for all of the nameless kinds that he came to know to exist before becoming aware of their existence. Before this awareness these kinds, and any names for them, were in no way objects of his thought.

This is especially apparent in texts such as *Meta.* 1.1 and *APo* 2.19. There Aristotle lists the stages of cognitive advance that culminate in knowledge of scientific principles, especially definitional principles. It is quite striking that knowledge of what names signify *as this is understood by supporters of the three-stage account* is not one of these stages. Nor is knowledge of what it would be to belong to a kind, in advance of knowledge of its existence, one of these stages. The initial stages are perception, memory and, then, the frequently repeated perception and collective memory of similar instances of the same type of thing that constitutes experience (ἐμπειρία).[10] As an example of accumulated experience concerning a kind, in *Meta.* 1.1, Aristotle offers the knowledge "that fire is hot" (981b13). This empirical knowledge obviously presupposes, indeed incorporates, knowledge of the existence of fire. Aristotle lists it as a starting point for inquiring *why* fire is hot and, thereby for him, inquiring for knowledge of the ultimate definitional principle as to what fire is.

[10] *APo* 2.19, 100a3ff, and *Meta.* 1.1, 980a27ff.

But there is no textual space on Aristotle's account for knowledge of what it would be to be fire *prior* to acquiring this empirical knowledge of its existence or *prior* to acquiring knowledge of this regular explicable feature of it. Whether there is room in *Meta.* 1.1 (and in *APo* 2.19) for a role for the grasp of nominal definitions on *another* account of their character than that offered in the three-stage view and others of its type, as there must be if they are to play a role in the advance to knowledge of definitional principles described there, we shall see later in this chapter.

A consideration of some of the semantic implications of the three-stage view helps to sharpen the issues here. On the three-stage account, for inquiring scientists to come to know of the existence of, say, thunder, according to Aristotle, they must know in advance of knowing the ultimate essence of thunder, in virtue of grasping the meaning of the name "thunder," a *necessary* condition for the existence of any actual thunder, a *general* property that must belong to every member of that natural kind "if" there are any. According to Aristotle himself, on this approach as described earlier, this necessary property in the case of thunder is in fact that of being "a noise in the clouds."[11] This implies that Aristotle means *to rule it out,* in advance of the discovery of the ultimate essence of the kind thunder, that thunder could occur anywhere else but in clouds. So he means to rule it out in advance, independently of knowing its ultimate cause and essence, that thunder could occur in a cloudless sky. This now would be poor semantics. Knowledge of the different forms that the real natural kind designated by our name "thunder" might take and where they may occur depends on the discovery and knowledge of the ultimate cause and essence which alone fixes the genuine boundaries of the kind. It cannot be known according to Aristotle himself, reasonably enough, before this (*Meta.* 8.4, 1044b9–15).

Connected with this, on this three-stage account, as noted, it is "sufficient" in order to be able to come to know of the existence and ultimate essence of, for instance, thunder to know of a "general" condition for the existence of thunder, namely that of being a noise in the clouds, a general condition which in fact fails to apply uniquely to thunder but does nevertheless give the ordinary signification of the name "thunder." On this view the name "thunder," according to Aristotle, given what it ordinarily signifies and designates, namely, generally, a noise in the clouds, *correctly applies,* for instance, to noises produced in clouds by tornadoes and noises produced in low hanging clouds by volcanoes. This idea is,

[11] Cf. *APo* 2.8, 93a22–23.

again, dubious semantics. The word "thunder" as we ordinarily use it does not correctly apply to these other types of noise, even to heard instances of these other types of noise that may be at the time phenomenally indistinguishable by us from certain known actual instances of thunder. On hearing these other noises one may still reasonably ask: was that thunder or not? Equally, a loss of light by the moon phenomenally indistinguishable by us from certain known instances of lunar eclipse would not count as a lunar eclipse, or thus be denoted by the term, should it be caused by the passage of a suitable meteorite between the earth and the moon not by the interposition of the earth between the moon and the sun.[12] So being phenomenally indistinguishable in such general terms from certain of these actual known instances is neither necessary nor sufficient for being denoted by the name in question or, thus, for being thunder or an eclipse as we use the terms. Moreover, to return to the issues in philosophy of science, meteorologists, in inquiring into the nature of thunder before knowing it, would hardly regard it as completely "sufficient" for guiding them to the knowledge of its nature simply to grasp a general feature of it that belongs equally to many other types of noise.

In sum, then, the three-stage account, as described by its proponents from Ross onward, is highly problematic, both as philosophy of science and as history of science, and as semantics as well. Nevertheless, this does not by itself show that it does not represent Aristotle's own views. However, since we have already found strong indication that he does not accept this account we should clearly exercise caution in reading those texts which are taken to confirm the attribution of it to him. There are, in fact, three main texts in *APo* 2 on which this attribution is based, in *APo* 2.1–2, in 2.8 and, finally, in 2.10. Let us consider these texts in turn.

Stages of Inquiry in *APo* 2.1–2

In *APo* 2.1 Aristotle begins by listing four types of questions which, he says, constitute the proper modes of search leading to forms of scientific knowledge. These are, first, the question of τὸ ὅτι, or of whether some fact obtains, which, says Aristotle, when we know it does, leads to the second question of τὸ διότι, or of why that fact obtains. Then there is a third question, εἰ ἔστι, or whether some kind of thing exists, which, when we

[12] *Pace*, for instance, Demoss and Devereux 1988, who propose that nominal definitions provide, in advance of knowledge of existence or essence, such sufficient, but not necessary, general conditions for kind membership.

know it does, leads to the fourth question, τί ἐστι, or what the essence of the existing kind is (89b23–35). Supporters of the three-stage view suppose that since Aristotle says here by way of example that one can inquire whether or not the centaur, or the moon, or the human species or night exists (89b32f, 90a5), he is presuming that one *must* already know what the name "centaur," or "moon," or "human being," or "night" signifies and that one must know by having that knowledge what it *would be* to be these kinds of things in advance, before and independently of coming to know whether there are such things.

But, however natural this reading might seem to be to some – we have already seen how unnatural it is in the cases of night or the moon – Aristotle does not say this or anything like it. He says nothing at all about names or their signification in *APo* 2.1–2 and nothing at all there about knowing in advance, in general terms, what it would be to belong to a kind *if* it exists; and his further analysis in *APo* 2.2 of what it is to pursue an inquiry as to whether something exists casts serious doubt on this presumption. At the beginning of *APo* 2.2 Aristotle continues his discussion of his four objects of search as follows:

> These [four] are the kinds of things we look for, and the things which we know when we discover them. However, when we look [either] for the fact (τὸ ὅτι) or for whether some thing exists simply (εἰ ἔστιν ἁπλῶς) we are [in both cases] looking for *whether there is a middle term for it;* and when we come to know either the fact or whether something exists [. . .] and we are then looking either for why the fact holds or for what the [existing] thing is, we are [in both cases] looking for *what the middle term is* [. . .]. It follows, then, that in all of our searches we are asking [not four questions but just two, namely] either whether there is a middle term or what the middle term is, *since* the middle term is the cause (αἴτιον), and in all these cases this is what is sought. (89b36–90a7)[13]

Thus Aristotle understands the question whether the moon exists to be closely comparable to the question whether it is a fact that the moon suffers eclipse. In the latter factual case the question concerns whether there are entities that regularly make up a lunar eclipse, such as the moon and a certain loss of light, entities whose connection is grasped as regular and so as warrantedly explicable through some middle term and cause that accounts for the connection between these two entities and thereby counts for Aristotle as what, ultimately, the thing is (90a5ff). As Aristotle indicates in *Phys* 2.8, what happens regularly in the same way either always or for the

[13] All translations are my own unless otherwise indicated.

most part happens, for him, in a natural and thus explicable way and can thereby be known as such before the explanation is found (198b34–199a7, b24–26). But the factual question: "Does the moon regularly undergo a certain loss of light?" does not use the name "eclipse," nor does one need to have a name for the eclipse to ask, or correctly answer, this question. The question whether the moon exists, then, also concerns in a parallel way whether there are features that make up the moon, such as, say, being a certain, nocturnally highly prominent, heavenly body, and waxing and waning in a certain manner, features whose connection is highly regular and thus, when grasped, warrantedly explicable through a middle term and cause for which one may search (90a9–11; cf. 90a14ff). So the question: "Does the moon exist?" is *reduced* by Aristotle to a question of fact such as: "Does a certain nocturnally prominent heavenly body wax and wane in a certain regular, and thus natural and explicable, manner?"[14] But to *ask* this latter question of fact one does not need to already have a name for the moon. This question does not use the name "moon." And to *answer* this question correctly in the affirmative one does not need first to know what it would be for there to be the moon while not "as yet" knowing of its existence in this manner. Rather, one can easily learn, as people including inquiring scientists no doubt did, what it would be, and is, for there to be the moon, in the sense in question, by virtue of learning or as the result of learning from perceptual experience that the nocturnally prominent nearby heavenly body in question waxes and wanes in a certain regular and thus warrantedly explicable manner, i.e., *by virtue of learning* of the existence of the moon as Aristotle understands it here, as a natural kind whose connected features are explicable, not necessarily before. And people can and likely did decide to give a special name to the moon only as or after they learned of its existence and what it is for it to exist in this manner not before.

Equally, to ask the question: "Does night exist?" is, for Aristotle, to ask a question such as: "Does the air around us lose its light between sunset and sunrise in a certain regular and thus explicable way?" To ask, and to correctly answer this question, one does not need to have – in advance of correctly answering it rather than by virtue of or as a result of correctly answering it – a name for night, or to already know in advance of this what it would be for there to be night. So Aristotle is not committed in

[14] Aristotle does not take it in *Phys* 2.8 that when there is some such regular connection this logically entails that it holds by nature, but he does take it that in such a case one is warranted in accepting it that it so holds and, thus, that one is in a position to search for its cause.

APo 2.1–2 to the highly implausible idea that *before* people, inquiring scientists or not, came to know of the existence of the moon or of night in the sense he has in mind here, they *must* have already had a name for the moon or night, and, simply by virtue of this, already had knowledge of what it would be to be these things, prior to and independent of knowing of their existence. In fact, even to ask, in the case of the moon, whether a certain nocturnally prominent heavenly body waxes and wanes in a certain regular manner, presupposes the existence of the body in question (with or without a name) in the ordinary sense.[15]

Some have presumed that in *APo* 2.1–2 and after Aristotle means to treat the question whether some kind *exists* in a radically different way than he does the question whether some *fact* obtains since, so they suppose, the question whether a kind exists arises for him only in the case of substances, such as the moon, while the question of fact is restricted to the case of attributes of substances, such as the eclipse.[16] If so, that might have a bearing on the issues in question here. But this is plainly false. The question of existence in *APo* 2.1–2, as we have just seen, is said to concern night just as it does the moon (90a5), and night is not a substance. It is rather, for Aristotle, just as is the lunar eclipse, a sort of periodic privation of light from a certain subject, namely a certain region of air where, so Aristotle says, this privation is due to the casting of a shadow by the earth on the thereby darkened region of air when the sun is appropriately positioned behind the earth in relation to that region of air.[17] Later, in *APo* 2.8, Aristotle applies the requirement to have suitable previous knowledge of *existence*, in order to discover ultimate cause and essence, alike to substances such as humankind and to attributes such as eclipse and thunder (93a20ff). In *Meta.* 7.17 too, the search for the nature and essence of humankind is treated as a search for why certain features of humankind are predicable one of the other, just as in the case there of the eclipse (1041a16, a32ff).

Some might fix on Aristotle's example in *APo* 2.1 of the centaur, as something about whose existence one might be inquiring, and argue that

[15] This curtails the use that can be made of *APo* 2.1–2 to show that for Aristotle a proper grasp of the ordinary meaning of the word "moon" or "night" does *not* involve an implicit presumption of or knowledge of their existence in the ordinary sense, if not necessarily their existence as natural kinds. The *American Heritage Dictionary*, it may be noted, defines the word "moon," in English, to mean "the natural satellite of the earth [...]" and the word "night" to mean "the period between sunset and sunrise [. . .]," where existence, even existence as a natural kind, is obviously presupposed and incorporated. These same terms are defined in the *Oxford English Dictionary* in the same manner, with even more correct empirical information built into their definitions there.

[16] Ross 1949, Bronstein 2016.

[17] *DA* 2.7, 418b18ff with *Mete.* 1.8, 345b8.

one needs the three-stage view to accommodate that example. But as we have seen from the cases of night and of the moon, even if that were plausible in this case one can hardly generalize to all cases from that example. Aristotle himself makes no such generalizing move in the text. In any event, moreover, as we have also just seen in the case of night, the object of inquiry concerning existence in the case of the centaur, as Aristotle explains it in *APo* 2.2, would be, say, whether there is a regular connection between the possession by certain animals of a horse-like lower body and a man-like upper body. This question is something one could pursue without having a name for any such animals and without possessing even a rudimentary account of or any commitment to what it would be in biological reality for there to be such animals, in terms of a general necessary (and/or sufficient) condition for belonging to such an actual species of animal, in advance of learning that there are such animals. A sensible scientist might well reserve judgment on what it would be, in biological reality, for there to be such animals, and reserve judgment on what property or properties might be biologically necessary (or sufficient) for being such animals *until* coming to knowledge of their ultimate essence, since, as we have seen for Aristotle, it is this essence that must determine the boundaries of the kind. In addition, assuming that there *are* centaurs, their existence could and likely would be discovered just as was the existence of the jaguar, by suitable experience of them, without requiring *in advance* either a name for them or knowledge of what it would be to be a centaur if they exist. So again, overall, what Aristotle says in *APo* 2.1–2 does not require, but rather carefully avoids, the most objectionable features that we have detailed of the three-stage view.

Supporters of that view have also often made use of Aristotle's discussion outside the *APo*, in *Phys* 4.6–9, of the void, which was posited as an unobservable theoretical entity by ancient Atomists to account for the possibility of motion. The Atomists themselves, however, did not introduce the term in question, with an understanding of what it would be for there to be the void, *before* coming to know of or to have sufficient reason, as they saw it, to posit its existence. Rather, they came to a first understanding of what it would be for there to be the void in the course of, as they supposed, discovering the need for and, thereby, the fact of its existence in order to explain motion, and then, after that, adapted an ordinary Greek term "the empty" (τὸ κενόν), giving it a special technical sense, as the name for it. (See, e.g., *Phys.* 4.7.) So the Atomists themselves were hardly committed to, nor did they use, the three-stage method in coming to know, as they supposed, of the existence and nature of the void.

Nevertheless, in discovering, as they did suppose, not only the existence of the void but also its nature and essence – as a spatial medium devoid in itself of all material body – the Atomists did commit themselves theoretically to their being this essential feature of the void as a necessary, and probably also a sufficient, condition of its existence. Thus, in an adversarial mode, Aristotle can properly argue, as he does against them, that there is no void by arguing that nothing could have this feature. But it hardly follows from this that Aristotle is committed to supposing that if the Atomists, or he himself, did *in fact* discover the existence of the void they or he would have *had* to do this by first having a name for it and/or by first having an account of what it would be to be the void while not yet having discovered or having sufficient reason to posit its existence. Aristotle's mode of rejection of the void requires no such objectionable general account of all scientific inquiry that leads to the *actual* discovery of the existence and nature of a kind, whether it be the void or any other. So what Aristotle says not only in *APo* 2.1–2 but also in related texts such as *Physics* 4.6–9 does not commit him to the three-stage account with its highly problematic features.

Stages of Inquiry in *APo* 2.8

The second main text which has been presumed to show most definitively that Aristotle is committed to the three-stage account is found in *APo* 2.8. There Aristotle is investigating just how one advances from proper knowledge of the existence of some kind to knowledge of its essence, a pattern of inquiry which he earlier identified already in 2.1–2. He now says, in line with what he said earlier, that in order to move from knowledge of existence to knowledge of essence the required knowledge of existence must be of a very special kind. Thus he says:

> Sometimes we grasp whether something *exists* accidentally, but sometimes when grasping *something of the thing itself* – for instance, [when we grasp] in the case of thunder, that there is a certain noise that belongs to clouds and, in the case of eclipse, that there is a certain loss of light [that belongs to the moon], and, in the case of humankind, that there is an animal of a certain sort [. . .]. In cases where we know [only] accidentally that something exists our grasp is necessarily in no way [directed] toward *what it is*. But in cases [of knowledge of existence] where we do grasp something [of the thing itself] it [our search for the essence] is much facilitated. (93a21–28)

Here by that kind of knowledge of existence which does facilitate a search for the essence, whether of substances such as humankind or of attributes such as eclipse, Aristotle clearly has in mind the very kind of knowledge of existence to which he assigned the same role in *APo* 2.1–2, namely the kind of knowledge of existence we have when we know in the case of some kind, "that there is a middle term for it," when we know, that is, that there are main features of the kind whose connection is regular and thus is, warrantedly, scientifically explicable through a middle term, even if we do not yet know what the middle term is. Here, then, also in *APo* 2.8, the knowledge of existence which Aristotle has in mind, which facilitates inquiry for the essence, is articulated knowledge of some fact concerning the kind which fact is understood to involve a regular connection and, thereby, to be a suitable object of scientific explanation; and, just as earlier in *APo* 2.1–2, nothing which Aristotle says in repeating this point requires that *before* one can come to such nonaccidental knowledge of existence, and independently of such knowledge of existence, one "must know" an account of what some word, some name for the kind, signifies and/or what it "would be" to belong to that very kind. Again, the statement of the fact the grasp of which counts as knowledge of existence in the case of the lunar eclipse – namely the fact that the moon periodically loses its light in a certain manner – does not even introduce the name "eclipse."

Nevertheless, some have argued that the special sort of knowledge of existence which Aristotle here has in mind, knowledge which we have when we grasp "something of the thing itself," or "something of what the thing is" as he later puts it (93a29), is not genuine knowledge of the actual existence of the kind, as earlier described, but rather is only grasp of a nominal definition known "at a stage prior to knowing that the kind exists."[18] However, this reading of *APo* 2.8 clearly involves a severe clash with the plain meaning of Aristotle's words. It requires us to suppose that when Aristotle speaks explicitly of that nonaccidental knowledge of the existence of some kind (which knowledge we have when we know "something of the thing itself") as involving one important way of "knowing that it [the kind] *exists*" (93a21ff), he does not mean what he actually says. On this reading Aristotle's special nonaccidental knowledge of existence is not knowledge of existence at all. This, it would clearly seem, is impossible. Nevertheless, it is argued, there is one subsequent passage in *APo* 2.8 that compels us to take this highly problematic line. There, according to the text adopted by Ross and followed by other supporters of the three-stage view, Aristotle says this:

[18] Charles 2000: 38; see also, e.g., Demoss and Devereux 1988, Bronstein 2016.

> Depending on *how* we grasp that something exists so are we directed [or not] to what it is. Of cases where we do grasp *something of what the thing is* [and *are* directed thereby to a final account of what it is by that non-accidental knowledge of its existence] let the first be thus: Let A stand for eclipse, C for moon, B for screening by the earth. To ask, then, *whether* (πότερον) the moon is eclipsed or not is to be searching for B, whether it exists or not. This is no different from asking whether there is an account of it. If it should be this we say that it too exists. (93a28–33)

Given this text, Aristotle can seem to say that we can already have nonaccidental knowledge of the *existence* of eclipses and still be searching for *whether* (πότερον) they exist. From this it might be inferred that nonaccidental knowledge of existence is not in fact knowledge of actual existence at all but rather is supplied in certain nominal definitions, or accounts of what it would be to be an eclipse if it exists, grasped prior to knowledge of actual existence. But that sharply conflicts with what Aristotle has just been saying about nonaccidental knowledge of existence, namely that it is an especially useful kind of *knowledge of existence* to have, as the basis for a search for the kind's essence not as a basis for the search for the kind's existence (93a20ff).

Fortunately, there is an alternate way of understanding this passage which avoids this major difficulty. This is based on an important alternate manuscript reading of Aristotle's text which is largely ignored by supporters of the three-stage account. On this alternate reading Aristotle says this:

> Depending on how we grasp that something exists so are we directed [or not] to what it is. Of cases where we do grasp *something of what the thing is* [and know already that it exists in that non-accidental manner] let the first be thus: Let A stand for eclipse, C for moon, B for screening by the earth. Then, to know the moon is eclipsed or not *in advance* (πρότερον) [by having already genuine non-accidental knowledge that it does undergo eclipse] is to be looking for B [screening by the earth, as middle term and cause], whether it exists or not. This is no different from looking for whether there is an account [or explanation] of it [i.e., of the moon's eclipse]. If it [i.e., the account] should be this [namely B, screening by the earth], we say that it too [B] exists [as the ultimate cause]. (93a29–33).

On this reading, which has at least as good manuscript support as the alternative printed by Ross,[19] Aristotle is not talking here at all about how we *acquire* knowledge of existence but only about how, having that in advance, in nonaccidental form, we acquire knowledge of essence. This fits

[19] Mss. ABd for πρότερον vs. Ms. n for πότερον.

perfectly with the tenor of his previous remarks where that is the only topic of discussion (93a20ff), and it does not severely clash with them as does the alternative reading. But on this more intelligible reading the passage has no implications for the issues we have been discussing concerning how one first comes to knowledge of existence or nonaccidental knowledge of existence; and so, on this reading of this passage, Aristotle does not commit himself here to the three-stage account.[20]

Stages of Inquiry in *APo* 2.10

The final main evidence offered in support of the attribution to Aristotle of the three-stage account comes in the famous passage at the beginning of *APo* 2.10. But here again there are serious textual issues. The heavily emended text offered by Ross and followed by other supporters of the three-stage account would be translated as follows:

> Since a definition is said to be an account of what something is, it is clear that one type of definition will be an account of what a name or other naming phrase signifies, e.g. what [the name] "triangle" signifies (οἷον τί σημαίνει τρίγωνον) – which very thing (ὅπερ) when we grasp (ἔχοντες) that it [the triangle] exists (ἔστι) we look for why it exists. But it is difficult in this manner to get a hold of it [of why it exists] if we do not [genuinely] know that it exists. The cause of the difficulty was stated earlier, namely that in this case we do not know whether the thing exists or not except accidentally. (93b29–35)

On this text Aristotle has been taken to commit himself to the three-stage view. First, we have a nominal definition and "when," i.e., after, that we aim for knowledge of existence; when we have that, we aim for knowledge of essence. However, this translation is in various ways questionable. First, no word for "when" (i.e., ὅτε), which might carry the suggestion that the grasping of the existence of something comes after the grasping of its nominal definition is found in the text. The text says simply: "Grasping (or in grasping, ἔχοντες) that it is we look for why it is." On this more neutral reading the text leaves it open that we may grasp that something exists in the course of coming to grasp a nominal account of what it is, not after. This latter possibility is strongly supported, moreover, by the common manuscript reading of the text, a reading which Ross discards, with little argument, as "impossible." On the common manuscript

[20] This analysis corrects the less attractive treatment of this passage offered in Bolton 1987.

reading, followed for instance by Bekker, Waitz, Detel, and Pellegrin, Aristotle rather says this:

> Since a definition is said to be an account of what something is, it is clear that one type of definition [or account of what something is] will be an account of what a name or other naming phrase signifies, for instance that [account of what something is] which signifies [i.e., indicates, in a preliminary way] what something is as a triangle (οἷον τὸ τί σημαίνει τί ἐστι ᾗ τρίγωνον) – which very thing (ὅπερ) [i.e., which very account], grasping that it is the case (ἔστι), we look for why it is the case. But it is difficult in this manner [by grasp of such a nominal account] to get a hold of this [of why that account is the case] if we do not [genuinely] know [in grasping the nominal account] that the thing exists, [since] in this event we do not know [in grasping the nominal account] whether the thing exists or not except accidentally. (93b30–35)

On this received manuscript reading the standard form for a nominal definition is not, for instance, as in Ross' emendation: "'Triangle' signifies such and such," but rather: "A triangle is such and such (as a triangle)." This fits appropriately with Aristotle's introduction of a nominal definition as one type of "account of what something is." The text, on the common manuscript reading, then, should be seen as intending to preserve this feature of nominal definitions, as initial accounts of what something is and, as such, accounts in which the name of the entity is not mentioned but only used. Once we follow the manuscripts, moreover, the natural antecedent for Aristotle's relative pronoun "which very thing" (ὅπερ) is not the triangle (which on Ross' reading is not even earlier mentioned – only the *word* "triangle" is). Rather the antecedent is that very thing which is the nominal *account* of what the triangle is, an account which uses but does not mention the word "triangle." This account is indeed what is understood by someone who has the word in their vocabulary, but there is no need for the word itself to be mentioned in the account.[21] On this reading, however, Aristotle is hardly supposing that the grasp of a nominal definition must be prior to or independent of knowledge of existence. Rather, as the passage directly indicates, he clearly is supposing that a nominal account *incorporates* at least one of the two forms of knowledge of existence which he distinguishes. The text leaves it open, moreover, that of the two possible types of existence-involving nominal accounts the only one that counts as a (preliminary) *definition* (e.g., of what something is *qua* triangle)

[21] Cf. Philoponus, *In APo* 372.18–19: "One who says: "A triangle is a figure with three angles." has uttered a nominal definition." Here the *word* "triangle" is not mentioned, but only used, in the nominal definition as Philoponus understands it.

is the one that incorporates nonaccidental knowledge of existence, as at *APo* 2.7, 92b4–34, where genuine definitions are so restricted.[22] Here, then, since we can make very good sense of what Aristotle says on the common reading of the manuscripts we have no choice but to prefer that reading over Ross' unnecessary emendation. On such a reading moreover, and crucially so, there is indeed now a place, unavailable on the three-stage account, for the grasp of nominal definitions in the sequence of cognitive stages that leads to the grasp of scientific principles in *Meta.* 1.1 and *APo* 2.19, namely at the stage of accumulated perceptual experience (ἐμπειρία) where one has a grasp of a nominal account as an explicable fact. Thus there is a very plausible, textually preferable, reading of this opening passage in *APo* 2.10 which does not introduce or support the three-stage account. So, in sum, none of the critical passages offered in support of this account requires us to attribute it, with all of its highly unattractive implications as history and philosophy of science, and as semantics, to Aristotle.

Aristotle's Procedures of Inquiry in *GA* 1

In *GA* 1 Aristotle offers us an especially clear example of the actual stages of inquiry which lead him to his account of the ultimate essence and definition of a kind in his treatment there of semen (σπέρμα).[23] So it will be useful here to proceed by focusing on that discussion. Does Aristotle there follow the three-stage procedure in his search? The opening and first stage of Aristotle's inquiry concerning σπέρμα is presented in *GA* 1.17 as follows:

> Some types of animals manifestly emit semen (σπέρμα), for example all whose nature it is to be blooded – it is unclear in which way the insects and cephalopods act. Therefore, we must investigate this: whether all males [blooded or otherwise] emit σπέρμα, or not all; and if not all what sort of cause is it by virtue of which some [males] do and others do not; and also

[22] Cf. *Meta.* 7.4, 1030b7ff.

[23] See, for discussion, Bolton 1987. There the main focus of attention is on the question whether, and to what extent, Aristotle uses his dialectical method in his inquiry as opposed to procedures offered in *APo* 2. Here the chief interest, though complimentary, is different. The new focus specifically concerns how Aristotle's procedures of inquiry in *GA* 1 fit, or not, the special mandates of the now popular three stage account. It is a major fault in typical presentations of this account and its close congeners that though its supporters suppose that this approach captures Aristotle's intended method in his scientific works they offer no text where Aristotle actually follows the stages to which they suppose he is committed in order to reach knowledge of the ultimate essence and definition of a kind.

> whether females contribute a kind of σπέρμα; and, if not σπέρμα, whether they contribute nothing else at all [to generation] or something but not σπέρμα. Then we must also ask what those who emit σπέρμα contribute through their σπέρμα to generation and, generally, what the nature of σπέρμα is, and also of the so-called menses in those animals which emit this fluid. (721a30–b6).

Here at this initial point in his inquiry, in saying that "some types of animals manifestly emit σπέρμα," Aristotle is clearly supposing that he knows *that* σπέρμα *exists*, and exists as a natural kind, as a contributing factor of some sort to natural generation which is made by certain males. So how did Aristotle come to this knowledge of its existence? There is no indication here or earlier in *GA*, or elsewhere, that he believed that in order to have come to this knowledge of its existence it was necessary for him, or for anyone, previously, while "not as yet" knowing of its existence, to have already had a name for σπέρμα, and a prior grasp of a general necessary, or sufficient, condition specifying what it would be to be σπέρμα if it exists which it was required to use to come first to know of its regular existence under certain conditions. As earlier with the cases of night and of the moon, the suggestion that this is how Aristotle himself, or others, came to knowledge of the existence of σπέρμα is rather ridiculous. He indicates here that its existence in certain familiar cases under certain regular well known circumstances is manifestly obvious. The most natural interpretation of this is that its existence under such regular circumstances in these familiar cases is simply perceptually obvious, which in fact of course it is. This fits well with Aristotle's direct claim in *Meta.* 1.1, as noted earlier, that:

> It is those with perceptual experience (ἔμπειροι) who know the *that*, but not the *why* [...]. The senses do not tell us the *why* in any case, *why* fire is hot for instance, but only *that* it is hot (981a28–b13).

As we have seen, knowledge of existence for Aristotle is a form of factual knowledge *that* which does not require having a name for the thing in question or previous knowledge of what it would be to be that thing before knowing of its existence by learning of it through suitable experience of that fact. Nor is the knowledge of the existence of σπέρμα which Aristotle evinces here in *GA* 1.17 based on the prior grasp of a general necessary feature of σπέρμα. Aristotle takes it as obvious that certain familiar males emit σπέρμα under certain well-known conditions but he does not yet know, as he makes very clear, whether it is necessary to be male, or to be a blooded male, to emit σπέρμα. For all he knows at this first stage, he says, females, or some females, may also contribute a type of σπέρμα in generation even though, clearly, if they do in the form of the *menses*, a possibility which remains open at this point, it would not have the phenomenal character of the most familiar cases in males. *A fortiori*, Aristotle does not offer here any general necessary *and* sufficient condition(s) for being σπέρμα, or, thus, any so-called commensurate universal concerning σπέρμα, which some would claim he requires for proper factual knowledge that a kind exists.[24]

One might perhaps argue that a passage such as this represents a juncture in Aristotle's inquiry concerning a kind which is *prior to* his employment of the three-stage account and that *subsequent to this* Aristotle means to first formulate a nonexistence-involving nominal definition as to what it would be to be σπέρμα if it exists, one that gives a general necessary condition for being σπέρμα, which he then employs to determine in a strict way that σπέρμα does exist prior to determining why it exists and what its ultimate cause and essence is.[25] But we find no such progression of thought in the text. After evincing the mode of knowledge of the existence of σπέρμα which we find initially expressed here in *GA* 1.17, Aristotle turns to a *causal* inquiry into the ultimate nature of σπέρμα. So he clearly supposes that his initial mode of knowledge of its existence, and existence as a natural kind, as expressed in *GA* 1.17, is fully sufficient for pursuing such a causal inquiry into its ultimate essence even though he does not yet know a general necessary, or necessary and sufficient, condition for its existence. Again, his knowledge of its existence is based simply on acquaintance in experience with the most familiar examples to us.[26]

Aristotle introduces his own causal inquiry in *GA* 1.18 as follows:

> The starting-point of this investigation [into the essence of σπέρμα], and of those that follow [into the menses, etc.], is first to grasp concerning σπέρμα what it is (τί ἐστι). For proceeding in this way we will be much better able to inquire into its functions (τὰ ἔργα) and into the derivative facts (τὰ συμβαίνοντα) about it. And σπέρμα is understood to be in its nature the sort of thing from which things which are naturally formed come primarily into being. (724a14–18)

[24] Lennox 1987, 1991.

[25] See Charles 2000, 6.2 and 10.6; but no text is offered where this pattern occurs. See also Bronstein 2016.

[26] This conforms to Aristotle's model for understanding the type of initial knowledge of existence that is incorporated in nominal definitions which is developed in Bolton 1976 and forthcoming, but not to the three stage account.

This passage seems to provide us with an initial or so-called nominal account of what σπέρμα is, as "the sort of thing from which things which are naturally formed come primarily into being." But it does not provide a nominal account of what σπέρμα is which Aristotle then goes on to use to establish its existence prior to looking for its essence. Rather it recalls Aristotle's earlier introduction and simply fixes on a feature of σπέρμα which was already presented as an obvious feature of well known cases of the actual existing thing – namely that it is, in certain instances at least, *one* sort of thing "from which" animals primarily come. This nominal feature of σπέρμα that we initially take certain manifest instances of σπέρμα to have, is then used not to establish the existence of σπέρμα, in some special mode, but rather as a given fact about σπέρμα which the proper basic causal account of its nature must explain. That is, this nominal fact as to what σπέρμα is gives Aristotle by itself enough to go on to begin right away to develop his own causal theoretical account of the nature of σπέρμα without any further question about its existence and without giving us what Aristotle takes to be a general necessary or sufficient condition of its existence. This feature is not, as we have seen, taken as a general necessary condition since the *menses*, at this juncture, may turn out to be σπέρμα but not satisfy this description, of being a primary source of generation. It is not a general sufficient condition since the *menses*, at this juncture, may turn out not to be σπέρμα but, for all we know, still satisfy this description. Similar points may easily be made concerning the possession, or not, of σπέρμα by male and/or female bloodless animals, about which, Aristotle says, he is not yet clear (*GA* 1.17, 721a31). Still, this nominal fact concerning what we take σπέρμα to be must incorporate already sufficiently strong nonaccidental knowledge *that σπέρμα exists*, for Aristotle, as the seminal material contributed by certain well-known cases of blooded males, to permit him to develop his own first account of its causal basis and essence. This he proceeds to do in *GA* 1.18, as follows:

> We must now say what the *primary nature* is of that which is called σπέρμα. It is necessary that everything which we find in the body [including σπέρμα] is either a natural part – and in that case either a non-homoeomerous or a homoeomerous part – or an unnatural part (such as a growth), or a residue, or a colliquation, or nourishment. (I call the surplus of nourishment *residue* and the secretion by unnatural decomposition from the material added in growth *colliquation*.) Clearly, σπέρμα cannot be a part, for although it is homoeomerous nothing is composed out of it as things are from sinew and flesh. Neither is it distinct, but all the remaining parts are. Nor is it unnatural or a deformity since it is originally present in

> all [its so far known possessors] and the nature of the [generated] thing comes into being from it. And nourishment is obviously imported from without. It is necessary therefore that σπέρμα is either a colliquation or a residue [. . .]. But a colliquation is something unnatural and nothing natural comes into being from what is unnatural. It is, therefore, necessary that σπέρμα is a residue. But every residue is either composed of useful or of useless nourishment. I call useless that from which nothing further is contributed to the thing's nature, and which is harmful if too much is used up, the opposite useful. That σπέρμα is not of the former sort is obvious. For those in the worst condition because of age or sickness have the most residue of that sort and the least σπέρμα [. . .]. Therefore, σπέρμα is a certain part of a useful residue; and the *final* residue is most useful and so that from which each of the parts comes directly into being. (724b21–725a13)

Here Aristotle introduces a complex division scheme to assist him in locating the cause and explanation of the chief fact already introduced earlier in his description of his mode of initial awareness that σπέρμα exists.[27] That fact, as it is recalled here, was that σπέρμα is "that from which each of the parts comes directly into being." That σπέρμα is the final residue of useful nourishment is established here by showing that it explains that fact. This fact is, of course, still quite nonspecific in various ways. Aristotle must still here be focusing on those familiar blooded males that he earlier took to be obvious contributors of σπέρμα and trying to explain why their contributions are generative. He goes on to make it very clear again that even at this first completed causal stage of inquiry he does not yet know whether females contribute a kind of σπέρμα or whether this explanation will apply to any female contribution. For all he knows here females, or some females, may contribute a type of σπέρμα but not something from which all of the parts of the generated animal come, or directly come in this manner, from the final residue of nourishment (*GA* 1.19, 726a29ff). These things only become clear as he further develops and deepens his causal story concerning animal generation. So Aristotle does not yet, even at this already completed strongly causal stage of inquiry, have a *general* account of what it would be to be σπέρμα, providing a general necessary (and/or sufficient) condition for being σπέρμα, something which, on the three-stage account, is required well prior to any causal stage. His procedures overall involve, rather, a first stage, reached by the grasp of a nominal account of what σπέρμα is that incorporates the

[27] See Bolton 1993, 2002; Lennox 1987; Charles 2000; and Bronstein 2016 for varying accounts of the uses of division by Aristotle in scientific inquiry.

accumulation of somewhat imprecise empirical data, or knowledge *that*, focusing on manifest cases, and a second stage, of causal inquiry, in the course of which those initial data come to be explained in a way that permits them and other data, eventually, to be refined and made properly precise. It is just these *two* stages of inquiry that Aristotle exhibits here in *GA* 1, just as he does in *APo* 2.1–2 where he reduces his initial four stages or modes of search for a scientist to just two – pursuit of the *that* and pursuit of the *why*.[28]

[28] For further articulation and defense of such a two stage approach, see Bolton and Code 2012 and Bolton forthcoming.

CHAPTER 13

Aristotle, Dissection, and Generation

Experience, Expertise, and the Practices of Knowing

James G. Lennox

Introduction

On Respiration begins with a complaint:

> A few of the earlier investigators of nature have discussed respiration; but for what reason it belongs to animals, some had nothing to offer, while others discussed it, but being insufficiently experienced with facts (ἀπειρ-οτέρως τῶν συμβαινόντων) did so poorly (*Juv.* 7 [= *Resp.* 1], 470b6–10).[1]

Aristotle's study of respiration[2] is ten Bekker pages long – four are devoted to a thrashing of his predecessors, in the midst of which he remarks:

> A most important cause of investigators not discussing these matters well is that they both lacked experience of the internal parts (τό τε τῶν μορίων ἀπείρους εἶναι τῶν ἐντός) and did not grasp that nature always acts for the sake of something; for by inquiring what respiration is present in animals for, and investigating this in connection with the parts, e.g. in connection with gills and lung, they would have discovered the cause sooner. (*Juv.* 9 [*Resp.* 3], 471b23–29)

Aristotle often attributes the failures of his predecessors to properly understand things to a lack of ἐμπειρία. As with this passage, however, it is often quite clear that the experience that would be needed to overcome this lack is not easy to come by – it requires *expertise*. In this essay, I will focus on a particular kind of expertise, and on what we can learn about Aristotle's understanding of it as a means to, and stage in, the acquisition of scientific knowledge. It is the extremely difficult and delicate expertise of the comparative anatomist – dissection. The vast majority of people who write

[1] Unless otherwise specified the translations are my own.

[2] Not actually a work. The translation of the opening lines suppresses its second word, γάρ. To quote Ross: "At 470b6, where Bekker marks the *De Respiratione* as beginning, there is no note, in any of our manuscripts, that a new work begins at this point, and the chapter begins with the words περὶ γὰρ ἀναπνοῆς, an impossible beginning for a separate treatise" (Ross 1955: 2).

about Aristotle appear to be unaware, or only vaguely aware, of how systematic and sophisticated his practice of dissection must have been. It is sometimes said, for example, that he needn't have dissected fish in order to write what he did about their internal anatomy – he just needed to wander down to the fishmonger's stall in the market, look around, and ask some questions. A comment in the *GA* about what fishermen believe based on experience without expertise is enough to cast serious doubt on that suggestion:

> The fact that the copulation of such fish is so rapid, so that it escapes the notice of many fishermen, also contributes to their being deceived; *for none of them ever keep watch over* (τηρέω) *such a thing for the sake of knowledge* (τοῦ γνῶναι χάριν). Nevertheless, copulation has been observed. Dolphins and these fish copulate in the same way, laying side by side; but the consummation in dolphins takes a longer time, while for such fish it is rapid. For which reason while not observing the coupling but seeing the swallowing of milt and eggs, the fishermen repeat over and over the same simple-minded tale about the conception of the fish, which Herodotus the fable-teller (μυθολόγος) also repeats, that fish conceive by swallowing the milt – not observing that this is impossible. For the passage whose entrance is through the mouth passes down into the stomach, not into the uterus; and whatever goes down into the stomach must of necessity become nourishment, because it undergoes concoction, while the uterus is apparently full of eggs – from whence did they arrive? (*GA* 3.5, 756a32–b13)

When Aristotle says that the fishermen do not watch over (τηρέω) such things for the sake of knowledge, it is helpful to read that in the wider context I have provided. To avoid the error to which the fishermen have fallen prey, one has to observe with a number of questions in mind, and a number of things already settled. Once you know the relevant internal anatomy, you *know* that the alternative to copulation that is being suggested is impossible, and you *know* by observation that the uteri of the female fish are full of eggs. Copulation is, based on wide-ranging observation, the most likely answer – so more careful, focused observation is needed, and when Aristotle writes "nevertheless, copulation has been observed," I take it he is reporting what he, or one of his colleagues, has observed *because they were observing for the sake of knowing*. That is, they observed with the eyes of the expert.

The problem is not, of course, that fishermen have no experience – good fishermen have just the sort of experience (and expertise) they need to locate and catch fish. What they do not have is the experience needed to dismiss the mistaken view about how the eggs of this particular species of fish are fertilized and incubated. That requires, among other things, a natural scientist's understanding of the reproductive anatomy and behavior of fish.

The purpose, then, of this paper is to explore the idea that the sort of experience that Aristotle often tells us was absent in the inquiries of his predecessors is experience, the acquisition of which requires domain specific expertise. On the face of it, this idea may seem to pose a conundrum. For if the person with the requisite expertise to make the appropriate observations – to have experience of the right kind – must already be, in some sense, a trained natural scientist, surely he would have to develop *that* expertise by means of some sort of prior, perception-based experience. And as we begin to travel down that road, an "experience/expertise" circularity seems to emerge. It will be best to explore this circularity further after we have Aristotle's thoughts on dissection-acquired experience in *GA*.

There are twenty-eight references to dissections in those works in the Aristotelian corpus assumed to be genuine, all but one in his animal investigations. Before tightening the focus to *GA*, I want to make some preliminary comments about the scientific use of diagrams in fourth-century BC Greece, and to Aristotle's use of diagrams more generally – because, while I will argue that not all twenty-eight references to "dissections" (ἀνατομαί) are to some sort of collection of diagrams, it is clear that some of these references are. And since the manuscripts transmitted to us do not include diagrams, and the collections of diagrams referred to in the ancient lists have been lost, a very serious problem arises: can we arrive at any sort of reasonable conjecture about what these diagrammatic images were like – their material character, their degree of abstraction, and their availability? More concretely, these references raise three quite different, though related, questions:

1. What is it that Aristotle is asking his readers to observe when he refers them to the dissections?
2. What sort of expertise was involved in their production, and what sort of expertise is required for the observations, either of diagrams or of the dissections they stand in for, to help you achieve knowledge?[3]
3. Do "the dissections" themselves represent a sort of knowledge – both for the expert who has produced them and for the student/investigator who is studying them – and if so, what sort of knowledge?

[3] To give you some idea of what I have in mind about expertise required to observe the diagrams, think of the difference between a skilled radiologist and an untrained patient looking at a set of x-rays. To "see" a bone fracture or a tumor in such images requires considerable training and skill.

In his *The Shaping of Deduction in Greek Mathematics*, Reviel Netz characterizes his first chapter as performing a trick, the trick of saying something substantial about the nonextant diagrams referred to in the manuscripts of Greek mathematical works. But, he notes, not all hope is lost.

> The texts – whose transmission is relatively well understood – refer to diagrams in various ways. On the basis of these references, observations concerning the practices of diagrams can be made. I thus start from the text, and from that basis study the diagrams. (Netz 1999: 12)

In a way, the investigation of Aristotle's many references to anatomical diagrams based on dissection needs to be even trickier, because in these references we are only rarely given hints about the things to which we are being referred. Aristotle rarely tells us directly what he thinks his readers should actually *do* when he says they should study something by means of the dissections. As we will see, the references to something *drawn* or *sketched* are few. Those few references might lead us to think of collections of drawings or diagrams to which students, or perhaps fellow researchers unfamiliar with the subject matter at hand, are being referred. But in their absence, this might be a recommendation to study the anatomical structure under consideration by *doing* dissections. And *that* is a very different recommendation, for performing a dissection in a way that actually allows you to visualize internal organs and their relationships to each other requires serious expertise. And even if the reader is being referred to diagrams, it is reasonable to assume that they are *based on* actual dissections performed by someone, and thus if we could discern anything about what these diagrams represent, and how they represent it, we might thereby learn something about the nature of the experience acquired by dissection, as practiced by those with the appropriate expertise, associated with the early Lyceum.

At any rate, my method here will be analogous to that of Netz: to consider the context in which Aristotle makes a recommendation to his readers to study something – typically a part or a system of parts – "in the dissections" or "by looking at the diagrams" or "the exemplars" in the dissections. Some common features of *all* of the texts in which these recommendations are made will first be noted, and I will then focus our attention on the texts in the *GA*. But first, some background on Aristotle's use of diagrams in general will be helpful.

Some (Nondissectional) Background

References to illustrations or diagrams – often lettered diagrams – are quite common in Aristotle's natural treatises outside the biological corpus. And

Reviel Netz reminds us that, for all we know, Aristotle was revolutionary in using and referring to diagrams in an investigation of nature.

> Our earliest direct evidence for the lettered diagram comes from outside mathematics proper, namely, from Aristotle. There are no obvious antecedents to Aristotle's practice. Furthermore, he remained an isolated phenomenon, even within the Peripatetic school which he founded. (Netz 1999: 61)

We can begin by looking at a number of such passages from *Meteorology* 1–3. I want to highlight two features of the language used in the references to diagrams in these passages that are shared with many of the references to the *Dissections* in the zoological treatises: [i] the use of the verb θεωρεῖν either in the imperative or in the infinitive with δεῖ; and [ii] the use of the preposition ἐκ with whatever it is that we are supposed to look at or study. The first feature is significant in suggesting, as Netz stresses about the use of mathematical diagrams, that Aristotle thinks there is a real cognitive *need*, in establishing certain conclusions or making certain points, to study these visual aids. The second is significant in suggesting a formulaic way of making these references, such that the use of the same formula in referring to "the dissections" may, with due caution, allow us to use these other passages as interpretive aids.

Mete. *1.8, 346a32–33 (on the Milky Way)*

> Let the circle as well as the stars within it be studied by means of the illustration. (θεωρείσθω [...] ἐκ τῆς ὑπογραφῆς)

This passage appears to be followed by a comment about it being impossible to position the "sporadic stars" on "the sphere" due to them having no fixed position. Sporadic stars are those not grouped into constellations. In any case, it is unclear whether Aristotle is referring to an *actual* sphere, or to some sort of two-dimensional *representation* of a sphere.[4] I have translated ὑπογραφή as "illustration" here because, unlike διάγραμμα, which has as one of its primary senses a figure marked out by lines, ὑπογραφή has a very wide range of usages, from legal documents to eye shadow! Aristotle uses ὑπογράψαντες at *GA* 2.6 to discuss the way in which nature behaves like painters who "drawn an outline" of the animal

[4] Thanks to email exchanges with Alan C. Bowen for helping me understand what I don't understand about this passage.

first and add the colors later (743b24), and in *GA* 4.1 the body consisting of flesh is said to be built around the blood vessels "as if around a framework" (ὡς περὶ ὑπογραφήν).

Mete. *2.5, 362a35–b20*

> For which reason those who currently draw the maps of the earth do so laughably (γελοίως γράφουσι νῦν...); for they draw the habitable world as a circle, but this is impossible both according to observations and according to reason (κατά τε τὰ φαινόμενα καὶ κατὰ τὸν λόγον). (362b13–15)

This passage follows a discussion of sectioning the earth into zones that are drum-shaped, the language of which is quasi-geometrical. He criticizes the current way in which "maps" (περίοδοι) of the earth are drawn. The depictions of the habitable world as a circle are said to be impossible for two reasons: first, "according to reason" because the earth is a sphere and has latitudes with determinate climates, so that there ought to be a continuous belt running all around the sphere that is habitable; second, "to the observations" of travelers who have provided reliable information about habitability that is inconsistent with the current maps. What is important here is not that Aristotle finds current "maps" (if that is the correct term) laughable, but that there is an identifiable geographic practice – craft? – of some sort that conveys information in the form of illustrations. Such illustrations can then be challenged both on theoretical grounds and by reliable observational reports. That means that while the use of illustrations of the internal organs of animals based on dissections to convey information about them to his readers might have been unprecedented in this particular field, it was not generally unprecedented in the science of his time.[5]

[5] Jacques Bromberg, my colleague in Classics, has pointed out the striking similarity between Aristotle's use of "map" (περίοδος) here and these passages in Herodotus:

"But I laugh (γελῶ) seeing how many before now have drawn (γράψαντας πολλούς) maps of the earth (γῆς περιόδους) and not one of them having any sense; for they draw Ocean flowing all around the earth, which is circular as if from a compass (κυκλοτερέα ὡς ἀπὸ τόρνου), and they make Asia and Europe equal [in size]" (*Hist.* 4.36.5–9).

"[...] as he [Anaxagoras] spoke he showed these things upon the map of the earth (ἐς τῆς γῆς τὴν περίοδον), which he had brought engraved upon a tablet (ἐν τῷ πίνακι ἐντετμημένην) [...]" (*Hist.* 5.49.24–25).

Cf. Aristophanes, *Clouds* (206–7): "This you have here is a map of the whole earth (γῆς περίοδος πάσης). Do you see? Here is Athens!"

Mete. *2.6, 363a26–364a4*

> Concerning the position [of the winds], it is necessary at the same time to study the accounts by means of the illustration (δεῖ δὲ περὶ τῆς θέσεως ἅμα τοὺς λόγους ἐκ τῆς ὑπογραφῆς θεωρεῖν). So then: the circle of the horizon has been drawn (γέγραπται) to provide more distinctness, wherefore it is also spherical. (363a26–28)

In this passage Aristotle makes extensive reference to an earth/globe diagram, and then uses the hypothetical imperative to posit a meaning for spatial opposition (363a30–33) and to associate letters with items in the diagram: "Let that to which A is applied be the equinoctial sunset, let that to which B is applied be its opposite [...]" (363a34–b7). This is a form of associating letters with elements in a diagram that is familiar in Euclid, and which we see in fragments of geometry before Euclid. Similarly, at *Mete.* 3.3, 373a1–18, during a discussion of lunar haloes, Aristotle uses phrases such as "Let AEB be joined" (373a10) and "The line drawn will therefore be a circle, and E the center" (373a16), clearly referring to some sort of lettered diagram. As we will see, Aristotle occasionally uses this form of imperative when he is directing his readers to features of diagrams in his zoological writings as well.

Mete. *3.5, 375b16–377a12 (on Rainbows)*

This geometrically rich and complicated passage is full of verbs for drawing, connecting points, etc. It opens with a claim that the main conclusions about the rainbow that Aristotle seeks to establish will actually be made clear by means of the diagram that is provided.

> Of the rainbow, that it is not able to be a circle or segment greater than a semicircle, and about its other attributes, will be clear by studying the diagram (ἐκ τοῦ διαγράμματος ἔσται θεωροῦσι δῆλον). (375b16–18)[6]

[6] Here is a partial list of other apparent references to lettered diagrams: *MA* 1, 698a18–b1; 9, 702b28–35; 11, 703b29–36; *IA* 7, 707b26–28; 9, 709a16ff.; 13, 712a1–13; *DC* 1.4, 271a7–23; 1.5, 272a7–21; 272b25–28; 1.6, 273a22–b27; 1.7, 275a14–b4; 2.4, 287b5–14. And, though not a reference to Aristotle's *own* diagrams, *DC* 1.10, 279b22–280a11 is interesting because in it Aristotle reflects on how geometers in the mid-fourth century BC understood the pedagogical role of diagrams. Aristotle is rejecting a "fictionalist" interpretation of the *Timaeus*, and tells how such an interpretation is typically characterized by its proponents – "alike, they say, to the drawing of diagrams; they speak about generation [of the cosmos] not because of a temporal coming to be but for the sake of teaching and as more familiar, just like watching the diagram [in geometry] come to be."

All of these passages provide valuable background and context for the references to dissections, to which we are about to turn, in a number of ways. First, as I will begin by noting, the linguistic parallels between these passages and the dissections references in the zoological treatises suggest that similar *cognitive needs* are being served, in making certain points clear and in aiding the reader to reach certain conclusions. Second, these passages show that Aristotle made widespread use of visual aids in the presentation of his research – perhaps he borrowed this technique from practices already common in geometry, or (as Netz remarks) perhaps Aristotle was the innovator, and the geometers were following his lead. In either case, this is not a technique that is limited to his zoological research, though perhaps having a distinct, and distinctly titled, set of books devoted to them might well have been. Finally, since the zoological passages so rarely mention diagrams, these passages from elsewhere in the corpus may supply some grounds for determining which passages are likely to be making such references even when diagrams are not explicitly mentioned.

The Language of the References to the *Dissections*

Diogenes Laertius 5.25 lists 8 books of *Dissections* and one referred to as *Selection from Dissections*. (The same two works are found in the list in the *Vita Hesychii*, but the *Dissections* is said to have only six books.) These lists are notoriously puzzling and unreliable, but in this case they provide presumptive evidence that there were materials, perhaps bound together in some way or in the form of scrolls, that were listed in catalogues of library holdings to which Diogenes had access. If Aristotle had not himself referred to such a collection so often, perhaps little more than that could be ascertained.

However, he did refer his readers to them regularly, and with a certain degree of consistency, and I would like to begin by noting some of the linguistic patterns that emerge regularly in these references. They provide valuable clues as to the role or roles Aristotle thought the practice of dissection and the study of its results played in the acquisition of knowledge about the natural world.

Distribution of Verbs

It will be noted that of the twenty-seven[7] verbs on the list, sixteen are explicitly normative in force, the normativity being conveyed either by

[7] Of those occurrences that TLG locates in works that are undisputedly by Aristotle, the only one I am *not* taking into account is that in *APo* 2.14. Though I think there are good reasons in favor of taking

	GA	*HA*	*PA*	*Resp.*	*Somn.*
δεῖ θεωρεῖν	2	1	4	1	
θεωρεῖν (imp.)		6	2		
θεωρεῖν (other)		2	1		
δῆλον (κατάδηλα)	3		1	1	
φανερόν		2			1

use of the imperative mood, by the use of verbal adjectives, or by the use of δεῖ with an infinitive. In fact, in virtually every case where Aristotle recommends that "the dissections" be observed, it is a *norm-governed* recommendation. Even when he is not explicit that there is a need to study something by means of the dissections, he claims that something will be *clearer*, *more apparent*, or *understood with more precision* by doing so.

Conjoint References to ἱστορίαι and ἀνατομαί

Among my chief concerns are to understand exactly what Aristotle thinks is achieved through "studying by means of dissections," why he thinks it is necessary that we do so, and whether there is, to borrow Netz' language "a mutual dependence of text and diagram"[8] (if diagrams there be) reflected in Aristotle's regular co-references to "histories" and "dissections." Outside of the *HA*, three of the six references in the *GA* are to both histories *and* dissections, as are seven of nine in the *PA*, and one of the two in *Resp.* Among the most interesting of these are those that *distinguish* the value of studying the one vs. the other: *GA* 2.7, 746a15; *PA* 4.5, 680a2; *Resp.* 16, 478b1. But it is possible that contextual differences between cases in which both are cited and those in which only one is (excluding *HA*, obviously) may well be instructive.

It is also worth noting, since translations sometimes mask this, that the manner of reference to "history" or "histories" is *also* quite variable: in *GA* and *Resp.* we are referred simply to "histories" in the plural, with no

it as a reference to animal dissections – all the examples in this chapter are zoological (including two (98a13–23) that report facts that could *only* be known by wide-ranging dissection), and the language of "selecting from dissections" (98a1–2) matches up with the reference to a work of selections from dissections in the Ancient lists – it is a disputed use of the term and, at any rate, is not in a zoological treatise.

[8] Netz 1999: 19–28.

explicit reference to animals; in *PA* there is one reference to "the natural history" in the singular (as at *IA* 1, 704b10), while the remainder refer to animals, but in a variety of ways.[9] In some cases, ἱστορία is in the singular and the reference to animals is in the plural, suggesting a single investigation of the differences of all the animals; while at other times it is in the plural, suggesting a collection of distinct investigations of animal differences.

Prepositions Linking the Verb to the Dissections *and* Histories

In most cases it is recommended that we study ἐκ the dissections, which I am tentatively treating as an instrumental use of ἐκ with the genitive. But there are a number of variants, and those that are most interesting are in the same passages in which Aristotle indicates that there are different values to be derived from studying histories and dissections. For in those cases the ἐκ goes with paradigms or diagrams that are said to be *in the Dissections*, and this is contrasted with what is *written* or *stated* in the histories (*GA* 2.7, 746a15; and compare *PA* 4.13, 696b15, *Resp.* 16, 478b1).[10] These passages may indicate that the expression "ἐκ τῶν ἀνατομῶν" is simply a condensed way of saying "by means of studying visual representations in collections of such representations known as "The Dissections." In any case, being alert to the differences among these references is likely to pay dividends.

Focusing on the *GA*

From this point on, I want to zero in on the passages in *GA*, and make use of references in the other biological works as aids in helping us make sense of these. In each case, I will first provide a context for the reference, provide text and translation for the reference, and follow this with a set of notes on the passage.

GA *1.10, 719a9–10*

Context

From 718b22–32, Aristotle discusses the vipers and selachia as examples of things that first produce an egg internally but are then externally

[9] The primary variants are: "the animal history" (668b30); "the history concerning the animals" (674b16, 689a19); "the histories concerning the animals" (680a2, 696b15).

[10] This form of reference to diagrams can also be seen in the *HA* (1.17, 497a32; 4.1, 525a8–9; 6.11, 566a14–15).

live-bearing. Then, between that discussion and the reference to dissections, he turns to a discussion of the uterus in these "ovovivipara," commenting that it is unlike either the ovipara or vivipara because it partakes in both of these forms, "for all selachians have the uterus next to the diaphragm and extending downward." He then goes on:

> Both about this and about the way in which the other uteruses are situated, one needs to have studied both by means of the dissections and the histories. (ἔκ τε τῶν ἀνατομῶν τεθεωπηκέναι καὶ τῶν ἱστοριῶν)[11]

Discussion

As I noted earlier, it is quite common for these passages to have a normative force, which comes either in the form, as here, of δεῖ with a form of θεῶρειν,[12] or a form of the imperative. The message appears to be that studying by means of dissections is something we need to do in order to fully understand the differences that he is characterizing among the uteruses found in the various kinds of blooded animals. After all, these are internal parts involved with generation, and as such require dissection to be observed. But this raises two of our initial questions – now in the context of an actual example – about what is being recommended, (the practice question) and to whom (the audience question). The form of the wording here leaves open at least two possibilities: Aristotle could be recommending that his audience perform dissections and look carefully at the reproductive organs; or he could be recommending that they go and study some drawings of various forms of uterus in a collection of illustrations, themselves based on observations resulting from dissections. In the *Meteorology*, Aristotle refers to illustrations, quite often lettered illustrations, on a number of occasions; and in a small number of cases, one of which we will look at shortly, he refers to some form of diagram in the *Dissections* as well.

That, however, only pushes our two questions back one step. For if there are collections of such illustrations to which his audience is being asked to refer, someone had to *produce* those illustrations, and to have any

[11] Though I am not focused on what "the histories" may refer to here, there are good reasons to be cautious about supposing that such references are to the very books that have come down to us, though presumably they bear some relationship to these references. The differences in the uteruses of the live-bearing, egg-laying and ovoviviparous animals are discussed in *HA* 3.1, 510b7–511a34, and the latter group is in focus at 510b34–511a18. It is of some interest that there is a reference to a lettered diagram at 510a29–b4: ἐκ τῆς ὑπογραφῆς τῆσδε. The demonstrative pronoun indicates that it accompanied the written description, though it must be acknowledged that, while the demonstrative appears in the best mss., it does not appear it all of them.

[12] This reference is unique, however, in using the perfective form of the infinitive, τεθεωρηκέναι.

value they need to faithfully represent the results of *actual dissections* – thus the "name" Aristotle uses in referring to the collection. Was Aristotle involved in *performing* these dissections, or in the *production* of the illustrations? Would both have been the task of slaves or craftsmen of some sort, carried out under his supervision? Perhaps the following familiar sentences, from his encomium to the scientific study of animals in *PA* 1.5, give us clues to the answers.

> For even in the study of animals disagreeable to perception, the nature that crafted them likewise *provides extraordinary pleasures to those who are able to know their causes* and are by nature philosophers. (645a8–10)

> For this reason, *we should not be childishly disgusted at the examination of the less valuable animals*. For in all natural things there is something marvelous. (645a15–17)

> If someone has considered the study of the other animals to lack value, he ought to think the same thing about himself as well; for *it is impossible to look at that from which mankind has been constituted – blood, flesh, bones, blood vessels, and other such parts – without considerable disgust*. (645a25–30)

The first two sentences might be taken simply to be referring to the unpleasant visual appearance of certain animals – though when one considers what it means for Aristotle to make a *causal* investigation of an animal, that suggestion is less likely. The third passage, however, stresses studying parts, and mentions blood vessels and bones in particular, which one does not study systematically without dissection. It would appear he is using all his rhetorical skill to urge his *audience* to perform – or at least observe – *dissections*, not merely to the study of diagrams.

At 719a9–10, we are being told that we need to consult the dissections and histories to be informed about an *extremely wide* inductive generalization: we are being asked to consider the nature of the uterus of all selachian fish in comparison with all other uteruses. That is, this reference to anatomical dissections is both about uteruses in general and about the *differences* between selachian uteruses and all other uteruses – and that may be precisely why one needs to study these things visually to get the point. To be precise, Aristotle is characterizing the nature of the uteruses of the ovoviviparous fish by locating them as intermediate in form between those of vivipara and of ovipara. Thus, though there is no direct reference to diagrams in this passage, its context and purpose strongly suggests that there are illustrations to which the reader is being referred, and that they must be quite abstract, intended to represent both what is common to all uteruses of a certain kind; and perhaps to highlight the differences between the kinds.

GA *2.4, 740a19–24*

Context

Our next reference is in the midst of Aristotle's argument that the heart is, as the source of both the uniform and nonuniform parts, the first part to be apparent in a clearly defined way "in all blooded animals." He goes on:

> For it is already appropriate to name the source (ἀρχή) of the animal and the system [of parts], once they have need of nourishment; for blood and its analogue are the last stage of nourishment, and the blood vessels are the receptacles of blood and its analogue, wherefore the heart is a source (ἀρχή) of the blood vessels. But this is clear from the histories and the dissections (δῆλον δὲ τοῦτο ἐκ τῶν ἱστοριῶν καὶ τῶν ἀνατομῶν). (740a23–4)

Discussion

This passage is one of the few references to the histories and dissections with no explicit normative tone and where the verb is not θεῶρειν. As with our first passage, the claim that is to be clarified by means of the Dissections and Histories is at a high level of generality – it is at least a claim about all blooded animals, but given Aristotle's reference to the analogue of blood, perhaps it extends to the entire animal kingdom. Moreover, since it is a claim about the heart as the source of the blood vessels in all animals, it is establishing a fundamental principle of his theory of generation. It is this system that distributes both the nutrients needed for the growth of each of the organs and the generative power that carries the "information" about the timing and character of development of the uniform parts that constitute the nonuniform parts, and he is here claiming that certain observations establish that the heart is the source of that system. It is thus important to consider the extension of τοῦτο in δῆλον δὲ τοῦτο.

There are good reasons to believe that it is referring to the fact that the blood vessels are the receptacle for blood and that the heart is their origin. Why would this be in need of support from the dissections? Because of a point Aristotle makes repeatedly in arguing for the heart's function: it is by a comparative anatomical study of many different internal organs and their connections to the blood vessels that you learn that the blood vessels don't "irrigate" the heart via a system of smaller and smaller vessels, as they do with other organs. Rather, they are connected to the heart as tubes that run directly from and to cavities within the heart. This proves that the blood vessels originate there – and only dissection (indeed, carefully performed dissection) can make this fundamental difference between the heart and all other visceral organs clear. This same point is stressed at *Resp.* 8, 474b7–9, where Aristotle insists that "the heart is the source of the blood vessels in blooded animals; for as it happens they [the blood vessels] do not run through the heart but from it, a fact that is clear (δῆλον) from the dissections." And in *PA* 3.4 on the heart, he concludes his argument for this very point by telling us that "these things are more clear (κατάδηλα μᾶλλον) by means of the dissections and the generations; for it is straightaway constituted of blood and first of all the parts to come to be" (669a9–11). Here, the γὰρ clause is meant to justify the unusual conjunction of a reference to the study of generation with the dissections; two things you learn by studying animal generation closely is that the heart is constituted directly from blood and that it is the very first organ to appear.

Regarding the question of the relationship between references to collections of visual representations of some sort and the activity of dissection that provides the inductive basis for those illustrations, on its own this passage is of little help. But the two supporting references in *PA* 3.4 and *Resp.* 8 that I just cited appear to be claims about what dissection itself actually shows, and are not references to collections of images based on dissections. The differences between the reference in passage 2 and the others in *GA* point in the same direction – Aristotle is here telling us what is clear from dissections themselves, not referring the reader to collections of representations.

GA *2.7, 746a9–18*

Context

This passage raises a number of special issues, and will require considerably more context setting that the first two passages. *GA* 2.1–5 has discussed in detail the nature of the causal contributions of male and female to generation: what happens when the capacity of the male, a capacity of the heat of the *pneuma* carried by the semen, initiates the process of development within the female material, up to the point where the heart becomes visible, and the heart then serves as the first principle for the ordered development of the body (740a8–23; 741b15–18) by perfecting the blood and by being the origin of the system of blood vessels by means of which nutrients and the generative principle are distributed to the developing embryo. Chapter 6 thus begins: "After the principle has come to be, as was said, the internal parts come to be prior to the external parts" (741b5). Aristotle then interweaves a complex discussion of the proper *method* to be used in studying development and the outlines of an account of the early stages of development. Our passage appears in chapter 7, a discussion of the means by which the embryo is nourished in live-bearing animals. It is unclear what subclass of organisms Aristotle has in mind in our passage, but the relevant contrast class consists of large and medium-sized live-bearing animals, including those lacking a complete set of upper teeth (i.e., ungulates), in which nourishment reaches the uterus by more than one blood vessel and the blood is then distributed to the umbilicus by means of "cotyledons," which, quoting Peck, "Aristotle uses [...] to mean the pits in the modified wall of the uterus into which the [chorionic] villi of the outer membrane of the embryo fit" (Peck 1942: 238 note b). Here is the passage with two references to *Dissections* embedded:

> Many of those animals that are stunted and have both upper and lower teeth do not have cotyledons in their uteruses, but the umbilicus extends to a single blood vessel that is large and stretches through the uterus. Although some animals of this sort produce one offspring and some more than one, those with multiple embryos have the same arrangement as those with one. *But it is necessary to study these things both by means of the exemplars in the dissections and the things written in the histories* (δεῖ δὲ ταῦτα θεωρεῖν ἔκ τε τῶν παραδειγμάτων τῶν ἐν ταῖς ἀνατομαῖς καὶ τῶν ἐν ταῖς ἱστορίαις γεγραμμένων). For the animals grow naturally from the umbilicus, and the umbilicus from the blood vessel, [the animals being] arranged in order one after another, along the flowing blood vessel as if along a trough; and there are membranes and the *chorion* around each of the embryos. And those who say the young in the uteruses are nourished through suckling on a small bit of flesh speaking incorrectly; for the same thing would happen in the other animals, *but as it stands that is not what is observed (for it is easy to study this by dissections)* (νῦν δ' οὐ φαίνεται (θεωρῆσαι γὰρ τοῦτο ῥᾴδιον διὰ τῶν ἀνατομῶν)). (746a8–22)

Discussion

Not only is this the first passage we have studied that makes explicit reference to visual representations of some sort – here referred to as exemplars (παραδείγματα) – but it is also one of four passages that explicitly *distinguish* what is provided for study by the *Dissections* from what is provided by the *Histories*. It will help to have the other three before us:

> How the position of the heart is related to the gills one should study by means of the dissections in relation to visual appearance and by means of the histories in relation to accuracy (πρὸς μὲν τὴν ὄψιν ἐκ τῶν ἀνατομῶν δεῖ θεωρεῖν, πρὸς δ' ἀκρίβειαν ἐκ τῶν ἱστοριῶν) [...]. (*Resp.* 16, 478a35–b1)

> None of the internal parts [of females] differ [from males] except for the uteruses, the visual appearance of which is to be studied by means of the diagrams in the dissections (ὧν ἡ μὲν ὄψις θεωρείσθω ἐκ τῆς διαγραφῆς ἐν ταῖς ἀνατομαῖς) [...]. (*HA* 1.17, 497a32)

> All the other hard-shelled animals, as was noted, also have the mouth, the tongue-like part, the stomach and residual vent, though they differ in position and magnitude. *The manner in which each of them has each of these parts should be studied both by means of the histories and the dissections concerning the animals; for some of these things need to be clarified by an account, others rather by reference to visual appearance* (ὃν δὲ τρόπον ἔχει τούτων ἕκαστον, ἔκ τε τῶν ἱστοριῶν τῶν περὶ τὰ ζῷα θεωρείσθω καὶ ἀνατομῶν· τὰ μὲν γὰρ τῷ λογῷ τὰ δὲ πρὸς τὴν ὄψιν αὐτῶν σαφηνίζειν δεῖ μᾶλλον). (*PA* 4.5, 679b35–680a4)

In our *GA* passage there is no explicit use of the language of ὄψις, but in the concluding criticism there is an appeal to what we now observe. (More about that interesting use of an appeal to dissection in a moment.) In its objective use this term either refers to the thing seen or to some aspect of its visual appearance, and that might be the way to read the *HA* passage, since it is the ὄψις we are to study by way of a diagram in the *Dissections*. But the other two cases, in which there is a contrast between studying by means of dissections in relation to ὄψις and studying by means of histories in relation to accuracy or reasoning, πρὸς τὴν ὄψιν could simply mean "in relation to (or for the purpose of) vision" or "visualizing." But what are we "visualizing," and what is the nature of the contrast?

Let's consider the various references to diagrams throughout the zoological works. There are two categories. First, besides the one noted in this section (*HA* 1.17, 497a32), there are two other clear references in *HA* to diagrammatic representations in the *Dissections*: διαγραφή at *HA* 4.1, 525a9; τὰ διάγραμμα at *HA* 6.11, 566a15. But there are also passages such as the well-known one at *HA* 3.1, 510a29–35,[13] which appears to refer to a *lettered* diagram representing the internal reproductive organs of male, live-bearing, land animals, but which makes no reference to this diagram being in the *Dissections*: θεωρείσθω δε τὰ εἰρεμένα ταῦτα ἐκ τῆς ὑπογραφῆς τῆσδε (510a29–30). Because there *is* a reference to σχήματα to be studied in the *Dissections* less than a Bekker page later (3.1, 511a12–13), it is possible that this diagram may be in the *Dissections* without explicitly saying so; but I think the τῆσδε speaks strongly against

[13] Compare *PA* 4.9, 684b21–685a5, apparently referring to a lettered diagram of the digestive tract of mollusks.

that possibility. That question aside, what is interestingly different about this passage – as well as the reference to a diagram of the mollusk intestinal system in *PA* 4.9 – is that we are given instructions about *how to map* the lettered features of the diagram on to the part represented. In that respect they are much more like the quasi-geometric references in *Meteorology*, *MA*, *IA*, and *DC* with which we began, than they are to any of the references to what is in the *Dissections*, for these latter references *never* mention a lettered correspondence between illustrations and parts.

Should we suppose that διαγραφή, διάγραμμα, ὑπογραφή, and παραδειγμά, all refer to the same sorts of things? At this point, I have no reason to think otherwise, but I am also open to the possibility that there are subtle differences in either reference or connotation. There may, on the other hand, be some interesting inferences to be drawn from a switch that takes place in the prepositions that are used. Unlike the passages that simply refer the reader to the *Dissections*, here ἐκ is used with the visual representations, while those representations are said to be ἐν the *Dissections*. In these cases, then, there is a clear demarcation between the entities referred to as "Dissections" and what is found *within* them; and what is found within them are visual images of some sort or other, referred to by the four words discussed earlier. These may take the form of collections of "white boards" (λευκώματα) with images on them, or, more likely, papyrus sheets collected together in some way or other. But at least in these cases we can be reasonably sure Aristotle is not telling his audience to go and either perform or observe actual dissections – he is directing us to collections of diagrammatic representations based, one assumes, on previously performed dissections. That in turn implies an act of *abstraction*, an attempt to represent in a visual way what is universal or common to a wide range of varying particulars. And the range of variation is typically very wide indeed – differences and similarities across *all* the uteruses of *all* the ovovivipara, or *all* the hearts of *all* blooded animals. To capture this in an illustration or a diagram is an act of abstraction, of leaving out the inessential features and the inessential variations in the represented features. It is a way of visualizing what is essential to the way in which, say, *all* vivipara nourish their embryos.

And that takes me to one final point about this passage, its closing reference to dissections. You will notice that I have not translated it as if it were a reference to a document of some sort. In this passage Aristotle wishes to give grounds for rejecting a claim made by someone that children in the womb derive nutrition by suckling on a bit of flesh. He does so by making the assumption that if this were true it would be true for all the live-bearing animals (or perhaps some subset of those that are appropriately similar to human beings), and then asserting that it can be easily seen διὰ τῶν ἀνατομῶν that this is not the case. Unlike the reference a few lines earlier, it is natural, especially given the different preposition, to see this as a reference to what one actually sees, or would see, when *performing, or at any rate watching,* dissections, rather than as a reference to a collection of diagrams.

In that case, in this passage Aristotle is making two theoretically interesting moves. One cannot directly observe how a human embryo is being nourished in the womb, and so there are questions about how one either argues against a speculation about how it happens or supports one's own view about it. This very passage shows us that Aristotle countenances important differences among live-bearing animals, so he cannot simply argue that a certain theory must be wrong because it violates the principle that all live-bearing animals nourish there young in precisely the same way. However, he *may* think the evidence from dissection supports the universal that all live-bearing animals provide nutrition to their young via an umbilical cord. Human children are born attached to an umbilical cord and this argues strongly in favor of them sharing that means of nutrition with other live-bearing animals. This then shows Aristotle appealing to the results of wide-ranging dissections not only in support of his own account of how fetuses are provided with nutrition, but in order to argue *against* a view about how *human* fetuses do so.

GA *4.1, 764a33–b4*

Context

The context here is a discussion of the explanations of Democritus and Empedocles for the fact that, in many cases of multiple births, some offspring are male and others female. He praises Democritus for at least seeking for the difference in the generation of the one vs. the other, though whether he has stated things well or poorly is "another *logos*" (764a23). He then considers the view, which he attributes to Empedocles (764a12),[14] that difference in temperature causes male parts to form in some and female parts to form in others. After noting a couple of "logical" problems in the case for this view, our passage begins.

> Again, female and male twins often come to be at the same time in the same part of the uterus, *and we have studied this sufficiently by means of the*

[14] An assumption Empedocles made in a "really more careless manner" (ὡς ἀληθῶς [...] ῥαθυμότερον, a12). Peck renders the comparative "rather slipshod," but I think Aristotle is intending an unfavorable comparison with the way Democritus approached the question.

dissections in all the live-bearing animals, both in those with legs and in the fish (καὶ τοῦθ᾽ ἱκανῶς τεθεωρήκαμεν ἐκ τῶν ἀνατομῶν ἐν πᾶσι τοῖς ζῳοτοκοῦσι, καὶ ἐν τοῖς πεζοῖς καὶ ἐν τοῖς ἰχθύσιν); about which things, if he [Empedocles] had not seen them, stating this [the temperature of the uterus] as the cause [of whether the fetus is male or female] would be a reasonable mistake, while if he had seen it, it would be absurd to think the cause is the heat or coldness of the uterus. For in that case both twins would become either female or male, but as it is (νῦν δὲ) we do not see (οὐχ ὁρῶμεν) this occurring.

Discussion

In this passage the reference to dissections has the usual verb (θεωρεῖν), but most interestingly a perfect first person plural, suggesting that Aristotle himself was the, or one of the, observers. That it is modified by the adverb ἱκανῶς suggests that Aristotle has views about the number, but also, given what he says *is* sufficient, the *range*, of dissections that are needed to either establish or refute a hypothesis. And when we consider the question of the range of observations he is discussing, how are we to understand the reference to live-bearing fish? I can only think of two possibilities: either ἰχθύς is being used very loosely here, to cover cetaceans; or he is using ζῳοτόκα very loosely to include ovoviviparous fish, the selachians. I incline to the latter alternative.

The first person plural verb, combined with the cataloguing of the groups in which "we" have made observations, and to a lesser extent what he says about holding the wrong view being a reasonable mistake until you see evidence that makes it impossible, makes it likely that this is another reference to *actual dissections*, not to diagrams. This likelihood is reinforced by all the visual language in this passage – if Empedocles (see 764a2–7, a12–15) hadn't *seen* males and females arising at the same time in the same region of the uterus, this would be an understandable error; while if he had, it would be absurd to hold the view he does. But as a matter of fact (νῦν δὲ) *we* – i.e., we who have actually looked – do not *see* (οὐχ ὁρῶμεν) this happen.

GA *4.4, 771b27–33*

Context

As in *GA* 4.1, in this passage Aristotle is appealing to what is observable by dissection to refute what would otherwise be a plausible theory. This passage thus shows Aristotle providing a *standard* that any adequate theory in this domain must meet. *It must accommodate the evidence of dissection.* In fact the form of the argument is similar to the previous passage: many offspring appear to be produced from one act of coition. Aristotle's view of how the male's capacity is delivered to the female material provides no obvious answers to questions about why different kinds of animals produce different numbers of offspring of different sizes and so on. In our passage, Aristotle is appealing to the evidence from dissection to reject an alternative to the theory he is about to present:

> Now the claim that the [different] locations within the uterus attract seed and that it is on this account that the embryos become many, due to the plurality of places and the cotyledons not being one, comes to nothing; for often two embryos come to be in the same place in the uterus, and it is apparent in those that bear multiple offspring when they are full of embryos lying in a row. *And this is clear by means of the dissections* (τοῦτο δὲ δῆλον ἐκ τῶν ἀνατομῶν ἐστιν).

Discussion

Once again in this passage, Aristotle appears to be referring to what is clear when you look at the results of actual dissection, rather than to a collection of illustrations or diagrams. Unlike the majority of references to dissections, there is no overt imperative here – something is clear from dissections that makes this interesting theory false. But there is, as I noted earlier, an implicit *norm* being observed here – a theory about what goes on inside a living organism must be tested against the observations arrived at by dissection. Failure to do so will almost certainly lead to a false conclusion. Perhaps for that very reason, this is also one of the few passages that does not use a form of the verb θεωρεῖν, the infinitive nearly always governed by δεῖ. (The presentation of his own theory, which follows these critical remarks, is a wonderful example of his appeal to more and less variation in an explanatory context – but that is a topic for another time.)

GA *5.1, 779a4–11*

Context

GA 5.1 begins with a quite lengthy methodological discussion raised by the fact that this book will study the attributes (παθήματα) by which the parts of animals differ. It is not immediately obvious why this should be a matter for a treatise on animal reproduction and development, but it turns out that the things to be studied arise, or at least are different, at different points in life. Nor are these always to be explained teleologically, so there is a long discussion of which features are likely to require explanation by reference to what they are for and which are not. The first affection that is taken up after this introduction is over is sleep,[15] and the habits of sleep among the offspring of various kinds of animals. But prior to that there is an *aporia* in need of resolution: when animals are first born, are they asleep or awake? Various plausible reasons are given for both alternatives, which is, of course the source of the *aporia*. Our passage stands at the beginning of Aristotle's attempt to resolve the problem:

> So, then, it is necessary that animals sleep for a great deal of time on account of the imposition of nutrition and of the weight from the upper region (but we have said that the cause of sleep is such as this in other places); but nevertheless waking appears to take place even inside the mother *and this becomes clear in the dissections and in those hatched in eggs* (δῆλον δὲ γίνεται τοῦτο ἐν ταῖς ἀνατομαῖς καὶ ᾠοτοκουμένοις),[16] then they drop off to sleep immediately once more. It is for this reason too that after being born the offspring passes a great deal of time sleeping.

Discussion

The conjunction ἐν ταῖς ἀνατομαῖς καὶ ἐν τοῖς ᾠοτοκουμένοις is unique, and raises some interesting questions. It is, I suppose, just possible to read the καὶ epexegetically, so that it is dissection of animals produced from eggs that is being cited as evidence. But it is more likely that Aristotle is pointing his readers to two distinct lines of evidence: dissections, and young born in eggs. You do not need dissections to, for example, here the chicks chirping away just before they peck their way through their eggshells. However, you do need dissections to see that animals in the womb are sometimes awake – it need not be vivisections, since the young can remain alive for a brief period after the mother has died. But the mere fact of movement within the womb, for which dissection is not needed, is not enough to prove the animal is awake. At any rate, this thought – that for vivipara one needs to answer this question via dissection, whereas for ovipara you can look at the young inside the laid egg – gives a reasonable explanation for the conjunction. The reference to other places where this has been discussed is presumably to *Somn.* 3, 456b17ff.

[15] I note in passing that this is prima facie odd, because the book opens (778a16–17) by saying that what must now be studied are the affections by virtue of which *parts* differ, and while he discusses variations in the way animals sleep, and sleep is plausibly thought of as an affection, Aristotle never mentions it being an affection of a part or parts in this passage. However, one of his opening examples, immediately after mentioning that the subject to be investigated is the affections of parts, is variation in voice (high vs. low); so I take it the actual parts involved can be backgrounded when the affections are the primary focus of his attention.

[16] Though the majority of MSS. have τοῖς ᾠοτοκοῦσιν, MS. Z reads ᾠοτοκουμένοις, which Drossaart Lulofs prints. I am following his text.

Conclusion

I concluded the introduction to this paper by identifying three questions about which I hoped to make progress toward answering by looking at those passages in the *GA* in which Aristotle directs his readers to consider the results of "dissections." I shall conclude by returning to those questions with the results of this investigation in hand.

1. *What is it that Aristotle is asking his readers to observe when he refers them to the dissections?*

It has become clear that while some of these passages are referring readers to some sort of visible representations of what dissection reveals, other passages are equally clearly either recommending that the reader examine internal parts by means of dissection, or telling the reader what would actually be seen upon performing dissections. The details of the actual wording in each passage have to be taken into account, and even then it is not always clear. In the case of those passages that refer to representations, there is enough variation in the terms used for these references, and enough variation in the generality of the representation, that difficult questions are raised about the form that these representations might take. One possibility is that they are highly schematic, abstracting away from the differences that would actually be seen on dissecting different kinds of organisms; another possibility is that a more concrete representation is considered as an exemplar for a wider class, a "model organism," so to speak. Unlike two passages I discussed in the *HA* and *PA*, where the text refers to letters on an accompanying diagram (similar to references to maps and geometric diagrams in the *Mete.*, *DC*, *MA*, and *IA*), these references to "the diagrams in the dissections" give us no hint either of the abstractness of the diagram or of their material nature.

2. *What sort of expertise was involved in their production, and what sort of expertise is required for the observations, either of diagrams or of the dissections they stand in for, to help you achieve knowledge?*

By looking carefully at these passages, it becomes clear that the facts that Aristotle believes will become clear to his readers by consulting dissections (either directly or through representations) are of a highly general nature – facts about the uteruses of extensive groups like the ovoviviparous fish, or the placement of embryos in organisms that give birth to multiple young. This could be due to selective but wide-ranging dissections, or it might reflect Aristotle's confidence that he can generalize from dissections of a small number of organisms to what is true of the wider class to which it belongs. But from the full range of these passages throughout the zoological treatises, and from his frequent remarks about problems encountered in performing dissections and how to avoid them, it is clear that he was a skilled dissectionist. Unfortunately, Aristotle's references don't include any remarks from which we can draw conclusions about the relationship between the number and range of dissections performed and the generality of the conclusions he reaches. And without a better understanding than I have at the moment of what form the diagrammatic representations took, it is also unclear what sort of expertise one would need either to produce them or to "read" them (on analogy with "reading" an image produced by x-ray or MRI).

3. *Do "the dissections" themselves represent a sort of knowledge – both for the expert who has produced them and for the student/investigator who is studying them – and if so, what sort of knowledge?*

Broadly speaking, Aristotle's appeal to the results of dissection support two different sorts of claims: [i] positive support for generalizations about the internal anatomy of wide classes of organisms, sometimes of a comparative nature and [ii] evidence against claims made about some feature of internal anatomy by Aristotle's predecessors. In the first category there is every reason to think the study of the representations of the results of dissections, or of the dissections themselves, is intended to provide perceptual, or quasi-perceptual, grounding for the investigator's generalizations about internal anatomy. In the second category, Aristotle typically is not only rejecting erroneous claims of his predecessors – he is simultaneously correcting them by pointing out what dissection actually reveals about the case in point. Recall the passages from *Resp.* with which I began this essay. Those who have attempted to reach conclusions about internal anatomy without dissection, "being insufficiently experienced with facts (ἀπειροτέρως τῶν συμβαινόντων) spoke about them poorly" (*Juv.* 7 [= *Resp.* 1], 470b9–10). On the other hand "by inquiring what respiration is present in animals for, and *investigating this in connection with the parts, e.g. in connection with gills and lung,* they would have discovered the cause sooner" (*Juv.* 9 [= *Resp.* 3], 471b23–29).

One last caution is in order, however, which applies to many of the references to dissections not only in *GA* but also in *PA* and *PN*: they have the tone of recommendations to students, or at any rate to those who are aiming to acquire knowledge about animals – and they are often *joint* recommendations, to study both "the histories" *and* "the dissections." Aristotle typically does not tell us whether studying both of these provides mutual and complementary support for the same generalizations, or different sorts of support for different generalizations about the same kind of part (or animal). However, as I discussed earlier with reference to *GA* 2.7, 746a9–18, there are a small number of texts that explicitly distinguish the visual support provided by "the dissections" from the verbal descriptions provided by "the histories"; and these texts lend some support to the idea that Aristotle thinks that our knowledge of internal anatomy is more secure if we study both visual representations and linguistic descriptions of the subject under discussion. Unfortunately, what little we are told on this score is insufficient to say much more than that. At *Juv.* 22 [= *Resp.* 16], 478a35–b1 Aristotle distinguishes the *visual appearance* provided by dissection from the *accuracy* provided by the histories, suggesting, perhaps, a distinction between the vividness or immediacy of a visual representation compared with the greater detail provided by the description of the ἱστορία; while at *PA* 4.5, 680a3–4, he distinguishes between two ways matters can be clarified – by *an account* (λόγος) and by reference to *visual appearance* (ὄψις)." That suggests a different contrast – different aspects of the internal anatomy of hard-shelled animals (ὀστρακόδερμα) will be clarified with the aid of visual inspection and by verbal description. Had those who were responsible for saving the manuscripts of Aristotle taken that message to heart, much of the speculation in this essay would be unnecessary. In any case, about one theme of this essay there can be very little doubt: to gain access to the kind of ἐμπειρία that Aristotle insists we need to reach the truth about the internal parts of animals and especially about the process of animal generation, the expertise of the skilled anatomist is indispensable.

References

Allen, J. 2015. Aristotle on Chance as an Accidental Cause. In Leunissen 2015: 66–87.

Anagnostopoulos, G. 2009. *A Companion to Aristotle*. Wiley-Blackwell, Malden and Oxford.

Balme, D. M. 1962. ΓΕΝΟΣ and ΕΙΔΟΣ in Aristotle's Biology. In *Classical Quarterly* 12: 81–98.

1987a. Aristotle's Biology was not Essentialist. In Gotthelf and Lennox 1987: 291–312.

1987b. Teleology and Necessity. In Gotthelf and Lennox 1987: 275–85.

1990. Matter in Definition: A Reply to G. E. R. Lloyd. In Devereux and Pellegrin 1990: 49–54.

1992. *Aristotle: De Partibus Animalium I and De Generatione Animalium I (with passages from II.1–3)*, edited by A. Gotthelf. Clarendon Press, Oxford.

Barnes, J. 1985. *The Complete Works of Aristotle*. The Revised Oxford Translation (= ROT). Princeton University Press, Princeton.

1994. *Aristotle: Posterior Analytics*. Clarendon Press, Oxford.

Barthélémy-saint-Hilaire, J. 1887. *Aristote: traité de la génération des animaux*. Hachette, Paris.

Bayer, G. 1995. Definition through Demonstration: The Two Types of Syllogisms in *Posterior Analytics* II.8. In *Phronesis* 40: 241–64.

Bekker, I. 1831. *Aristotelis opera*. Academia Regia Borussica, Berlin.

Besnier, B. 1997. L'âme végétative selon Aristote. In *Kairos* 9: 33–56.

Beullens, P. and A. Gotthelf. 2007. Theodore Gaza's Translation of Aristotle's *De Animalibus*: Content, Influence and Date. In *Greek, Roman, and Byzantine Studies* 47: 469–513.

Bolton, R. 1976. Essentialism and Semantic Theory in Aristotle: *Posterior Analytics* II.7-10. In *Philosophical Review* 85: 514–44 (repr. in Bolton 2010b: 463–96).

1987. Definition and Scientific Method in Aristotle's *Posterior Analytics* and *Generation of Animals*. In Gotthelf and Lennox 1987: 120–66 (repr. in Bolton 2010b: 163–221).

1993. Division, définition et essence dans la science aristotélicienne. In *Revue philosophique* 2: 197–222 (repr. in Bolton 2010b: 359–83).

2002. Substance and the Definition of Definition in Aristotle. In Canto-Sperber and Pellegrin 2002: 155–81.

2010a. Aristote et la signification des noms. In Bolton 2010b: 497–509.

2010b. *Science, dialectique et éthique chez Aristote*. Peeters, Louvain-La-Neuve.

2012. Science and Scientific Inquiry in Aristotle: A Platonic Provenance. In Schields 2012: 46–60.

Unpublished. *Techne* and *Empeiria*: Aristotle's Contribution to the Tradition.

Bolton, R., and A. Code. 2012. Aristotle on Knowledge. In Hetherington 2012: 50–71.

Bonitz, H. 1870. *Index aristotelicus*. Academia Regia Borussica, Berlin (reprinted Walter de Gruyter 1961).

Bos, A.-P. 2003. *The Soul and its Instrumental Body: A Reinterpretation of Aristotle's Philosophy of Living Nature*. Brill, Leiden and Boston.

Bowen, A. C. 1991. *Science and Philosophy in Classical Greece*. Garland, New York.

Bronstein, D. 2012. The Origin and Aim of *Posterior Analytics* II.19. In *Phronesis* 57: 29–62.

2016. *Aristotle on Knowledge and Learning: The Posterior Analytics*. Oxford University Press, Oxford.

Brunschwig, J. 2004. *On Generation and Corruption* I. 1: A False Start? In de Haas and Mansfeld 2004: 25–63.

Burnyeat, M. 2004a. Aristotle on the Foundations of Sublunary Physics. In de Haas and Mansfeld 2004: 7–24.

2004b. Aristotelian Revisions: The Case of the *de Sensu*. In *Apeiron* 37: 177–80.

Calow, P. 1976. *Biological Machines: A Cybernetic Approach to Life*. Edward Arnold Ltd., London.

Canto-Sperber, M., and P. Pellegrin. 2002. *Le style de la pensée: Recueil de textes en hommage à Jacques Brunschwig*. Les Belles Lettres, Paris.

Carbone, A. 2011. *Aristote illustré: représentations du corps et schématisation dans la biologie aristotélicienne*. Classiques Garnier, Paris.

Cerami, C. 2014. *Nature et sagesse: Les rapports entre physique et métaphysique dans la tradition aristotélicienne. Recueil de textes en hommage à Pierre Pellegrin*. Peeters, Louvain-La-Neuve.

2015. *Génération et substance: Aristote et Averroès entre physique et métaphysique*. Walter de Gruyter, Berlin and New York.

Charles, D. 1990. Aristotle on Meaning, Natural Kinds and Natural History. Devereux and Pellegrin 1990: 145–67.

2000. *Aristotle on Meaning and Essence*. Clarendon Press, Oxford.

Charles, D., and M. Frede. 2000. *Aristotle's Metaphysics Lambda: Symposium Aristotelicum*. Oxford University Press, Oxford.

Clearly, J. J., and D. C. Shartin. 1988. *Proceedings of the Boston Colloquium in Ancient Philosophy* 4.

Code, A. 1997. The Priority of Final over Efficient Cause in Aristotle's Parts of Animals. In Kullmann and Föllinger 1997: 127–43.

Connell, S. 2016. *Aristotle on Female Animals: A Study of the Generation of Animals*. Cambridge University Press, Cambridge.
Cook, K. C. 1996. Sexual Inequality in Aristotle's Theories of Reproduction and Inheritance. In Ward 1996: 51–67.
Cooper, J. M. 1990. Metaphysics in Aristotle's Embryology. In Devereux and Pellegrin 1990: 55–84.
Coughlin, S. *On the Principle of Separation in Aristotle's Biology*. Unpublished.
de Haas, F. A. J., M. Leunissen, and M. Martin. 2010. *Interpreting Aristotle's Posterior Analytics in late Antiquity and the Byzantine Period*. Brill, Leiden.
de Haas, F. A. J., and J. Mansfeld. 2004. *Aristotle's on Generation and Corruption: Book I*. Symposium Aristotelicum. Clarendon Press, Oxford.
Demoss, D., and D. Devereux. 1988. Existence, Essence and Nominal Definition in Aristotle's *Posterior Analytics* II.8–10. In *Phronesis* 33: 133–54.
De Raeymaeker, L. 1955. *Autour d'Aristote: recueil d'études de philosophie ancienne et médiévale offert à Monseigneur A. Mansion*. Publications Universitaires de Louvain, Louvain.
Deslauriers, M. 1998. Sex and Essence in Aristotle's Metaphysics and Biology. In Freeland 1998: 138–67.
Detel, W. 1993. *Aristoteles, Analytica Posteriora*. Akademie Verlag, Berlin.
Devereux, D., and P. Pellegrin. 1990. *Biologie, logique et métaphysique chez Aristote*. Éditions du CNRS, Paris.
Dixsaut, M., A. Castel-Bouchouchi and G. Kévorkian. 2013. *Lectures de Platon*. Ellipses, Paris: 215–31.
Drossaart Lulofs, H.J. 1975. *Aristotelis De generatione animalium*. Clarendon Press, Oxford.
Falcon, A. Aristotle on the Study of Animals and Plants. In Holmes and Fischer 2015: 75–91.
Fine, G. 2014. *The Possibility of Inquiry. Meno's Paradox from Socrates to Sextus*. Oxford University Press, Oxford.
Freudenthal, G. 1995. *Aristotle's Theory of Material Substance, Heat and Pneuma, Form and Soul*. Oxford University Press, Oxford.
Freeland, C. A. 1998. *Feminist Interpretations of Aristotle*. Pennsylvania State University Press, University Park (PA).
Furth, M. 1988. *Substance, Form and Psyche: An Aristotelian Metaphysics*, Cambridge University Press, Cambridge.
Gelber, J. 2010. Form and Inheritance in Aristotle's Embryology. In *Oxford Studies in Ancient Philosophy* 39: 183–212.
Gill, L. M., and P. Pellegrin. 2006. *A Companion to Ancient Philosophy*. Wiley-Blackwell, Malden and Oxford.
Goldin, O. 1996. *Explaining an Eclipse: Aristotle's Posterior Analytics II.1–10*. University of Michigan Press, Ann Arbor.
Gotthelf, A. 1985. *Aristotle on Nature and Living Things*. Mathesis Publications, Pittsburgh (PA), and Bristol Classical Press, Bristol.

1985. Notes towards a Study of Substance and Essence in Aristotle's *Parts of Animals* II–IV. In Gotthelf 1985: 27–54.

1987. Aristotle's Conception of Final Causality. In Gotthelf and Lennox 1987: 226–54 (repr. in Gotthelf 2012a: 3–44).

1988. The Place of the Good in Aristotle's Teleology. In Clearly and Shartin 1988: 133–39 (repr. in Gotthelf 2012a: 45–66).

1989. Teleology and Spontaneous Generation in Aristotle: A Discussion. In Penner and Kraut 1989: 181–93 (repr. in Gotthelf 2012a: 142–50).

2012. Teleology and Embryogenesis in Aristotle's *Generation of Animals* II 6. In Gotthelf 2012a: 90–116.

2012a. *Teleology, First Principles, and Scientific Method in Aristotle's Biology*. Oxford University Press, Oxford.

Gotthelf, A., and J. G. Lennox. 1987. *Philosophical Issues in Aristotle's Biology*. Cambridge University Press, Cambridge.

Hall, B. K., and W. Olson 2003. *Keywords and Concepts in Evolutionary Developmental Biology*. Harvard University Press, Cambridge (Mass.).

Hayduck, M. 1903. John Philoponus [= Michael of Ephesus] *In libros de generatione animalium commentaria*. In *Commentaria in Aristotelem Graeca* 14.3. Reimer, Berlin.

Henry, D. 2005. Embryological Models in Ancient Philosophy. In *Phronesis* 50: 1–42.

2006a. Understanding Aristotle's Reproductive Hylomorphism. In *Apeiron* 39: 269–89.

2006b. Aristotle on the Mechanisms of Inheritance. In *Journal of the History of Biology* 39: 425–55.

2007. How Sexist Is Aristotle's Developmental Biology? In *Phronesis* 52: 251–69.

2009. Aristotle's *Generation of Animals*. In Anagnostopoulos 2009: 368–83.

2011. Aristotle's Pluralistic Realism. In *The Monist* 94: 197–220.

2013. Optimality and Teleology in Aristotle's Natural Science. In *Oxford Studies in Ancient Philosophy*: 225–63.

2015. Holding for the Most Part: The Demonstrability of Moral Facts. In Henry and Nielsen 2015: 169–89.

Henry, D., and K. M. Nielsen. 2015. *Bridging the Gap between Aristotle's Ethics and Science*. Cambridge University Press, Cambridge.

2015. Introduction. In Henry and Nielsen 2015: 1–25.

Hetherington, S. 2012. *Epistemology: The Key Thinkers*. Continuum, New York and London.

Holmes, B., and K. D. Fischer. 2015. *The Frontiers of Ancient Science: Essays in Honor of Heinrich von Staden*. Walter de Gruyter, Berlin.

Johnson, M. R. 2005. *Aristotle on Teleology*. Oxford University Press, Oxford.

Judson, L. 1991. Chance and "Always or For the Most Part" in Aristotle. In Judson 1991a: 43–100.

1991a. *Aristotle's Physics: A Collection of Essays*. Oxford University Press, Oxford.

Karbowski, J. 2012. Slaves, Women, and Aristotle's Natural Teleology. In *Ancient Philosophy* 32: 323–50.

Kosman, A. 2010. Male and Female in Aristotle's *Generation of Animals*. In Lennox and Bolton 2010: 147–67.

Kullmann, W. 1979. *Die Teleologie in der aristotelischen Biologie: Aristoteles als Zoologe, Embryologe und Genetiker*. Sitzungsberichte der Heidelberger Akademie der Wissenschaften. Phil-hist. Klasse Abhandlung 2. Winter, Heidelberg.

Kullmann, W., and S. Föllinger. 1997. *Aristotelische Biologie: Intentionen, Methode, Ergebnisse*. Franz Steiner Verlag, Stuttgart.

Kupreeva, I. 2010. Causation and Conditional Necessity in *Posterior Analytics* II 12. In de Haas and Leunissen 2010: 203–33.

Laks A., and M. Rashed. 2004. *Aristote et le mouvement des animaux: Dix études sur le De motu animalium*. Presses de l'Université de Lille, Éditions du Septentrion, Lille.

Le Blond, J.-M. 1939 (1996). *Logique et méthode chez Aristote*. Vrin, Paris.

Lefebvre, D. 2014. La jument de Pharsale. Retour sur *De generatione animalium* IV, 3. In Cerami 2014: 207–72.

2016. Le *sperma*: forme, matière ou les deux? Aristote critique de la double semence. In *Philosophie antique* 16: 31–62.

Lennox, J. G. 1987. Kinds, Forms of Kinds, and the More and the Less in Aristotle's Biology. In Gotthelf and Lennox 1987: 339–59.

1991. Between Data and Demonstration: The *Analytics* and the *Historia animalium*. In Bowen 1991: 261–95 (repr. in Lennox 2001b: 39–71).

1996. Aristotle's Biological Development: The Balme Hypothesis. In Wians 1996: 228–48.

2001a. Are Aristotelian Species Eternal? In Lennox 2001b: 131–59.

2001b. *Aristotle's Philosophy of Biology: Studies in the Origins of Life Science*. Cambridge University Press, Cambridge.

2001c. *Aristotle, On the Parts of Animals*. Translated with a Commentary by J. G. Lennox. Clarendon Press, Oxford.

2004. The Place of Zoology in Aristotle's Natural Philosophy. In Sharples 2004: 58–70.

2010. The Unity and Purpose of on *Parts of Animals* 1. In Lennox and Bolton 2010: 56–77.

2011. Norms of Inquiry. In *HOPOS: The Journal of the International Society for the History of Philosophy of Science* 1: 23–46.

Lennox, J. G., and R. Bolton. 2010. *Being, Nature, and Life in Aristotle*. Cambridge University Press, Cambridge.

Lesher, J. 2010. From Inquiry to Demonstrative Knowledge. *New Essays on Aristotle's Posterior Analytics. Special Issue of Apeiron: A Journal for Ancient Philosophy and Science* 43.

Leunissen, M. 2010. *Explanation and Teleology in Aristotle's Science of Nature*. Cambridge University Press, Cambridge.

2010a. Aristotle's Syllogistic Model of Knowledge and the Biological Sciences: Demonstrating Natural Processes. In Lesher 2010: 31–59.

2012. Aristotle on Natural Character and Its Implications for Moral Development. In *Journal of the History of Philosophy* 50: 507–30.

2013. "Crafting Natures": Aristotle on Animal Design. In *Philosophic Exchange* 41: 28–51.

2015. *Aristotle's Physics: A Critical Guide*. Cambridge University Press, Cambridge.

Leunissen. M., and A. Gotthelf. 2010. What's Teleology Got to Do with It? A Reinterpretation of Aristotle's *Generation of Animals* V. In *Phronesis* 55: 325–56 (repr. in Gotthelf 2012a: 117–41).

Lloyd, G.E.R. 1990. Aristotle's Zoology and His Metaphysics: The *status quaestionis*. A Critical Review of Some Recent Theories. In Devereux and Pellegrin 1990: 7–35.

1991. *Methods and Problems in Greek Science*. Cambridge University Press, Cambridge.

1996. *Aristotelian Explorations*. Cambridge University Press, Cambridge.

Louis, P. 1961. *Aristote: De la génération des animaux*. Les Belles Lettres, Paris.

Matthews, G. 1986. Gender and Essence in Aristotle. In *Australasian Journal of Philosophy*. Supplement to Vol. 64: 16–25.

Moraux, P. 1955. A propos du νοῦς θύραθεν chez Aristote. In L. de Raeymaker 1955: 255–95.

Müller G. B., and L. Olsson. 2003. Epigenesis and Epigenetics. In Hall and Olson 2003: 114–23.

Needham, J. 1934. *A History of Embryology*. Cambridge University Press, Cambridge.

Netz, R. 1999. *The Shaping of Deduction in Greek Mathematics*. Cambridge University Press, Cambridge.

Nielsen, K. 2008. The Private Parts of Animals: Aristotle on the Teleology of Sexual Difference. In *Phronesis* 53: 373–405.

Peck, A. L. 1942. *Aristotle: Generation of Animals*. Harvard University Press, Cambridge (Mass.).

Pellegrin, P. 1982. *La classification des animaux chez Aristote. Statut de la biologie et unité de l'aristotélisme*. Les Belles Lettres, Paris (Engl. trans. *Aristotle's Classification of Animals: Biology and the Conceptual Unity of the Aristotelian Corpus*. University of California Press, Berkeley and Los Angeles).

1986. Les fonctions explicatives de l'*Histoire des animaux* d'Aristote. In *Phronesis* 31: 148–66.

1990. Taxinomie, moriologie, division: Réponses à G. E. R. Lloyd. In Devereux and Pellegrin 1990: 37–48.

2005. *Aristote: Seconds analytiques*. Flammarion, Paris.

2006. Ancient Medicine and Its Contribution to the Philosophical Tradition. In Gill and Pellegrin 2006: 664–85.

2011. *Aristote, Les Parties des animaux. Traduction et présentation*. Flammarion, Paris.

Penner, T. and R. Kraut. 1989. Nature, Knowledge and Virtue. *Apeiron* 22.

Platt, A. 1912. De generatione animalium. In *The Works of Aristotle*, volume 5, edited by J. A. Smith and W. D. Ross. Clarendon Press, Oxford.

Potter, P. 2012. *Hippocrates: Generation, Nature of the Child, Diseases IV, Nature of Women, Barrenness*. Harvard University Press, Cambridge (Mass).

Rashed, M. 2002. La préservation (σωτηρία), objet des *Parva Naturalia* et ruse de la nature. In *Revue de philosophie ancienne* 20: 35–59.

2004. Agrégat de parties ou *vinculum substantiale* ? Sur une hésitation conceptuelle et textuelle du corpus aristotélicien. In Laks and Rashed 2004: 185–202.

2005. *Aristote, De la génération et la corruption.* Les Belles Lettres, Paris.

2013. Platon et les Mathématiques. In Dixsaut *et al.* 2013: 215–31.

Ross, W. D. 1949. *Aristotle: Prior and Posterior Analytics.* Clarendon Press, Oxford.

1955. *Aristotle: Parva Naturalia.* Clarendon Press, Oxford.

1958. *Aristotle: Metaphysics.* Clarendon Press, Oxford.

1961. *Aristotle: De Anima.* Clarendon Press, Oxford.

Schields, Ch. 2012. *The Oxford Handbook of Aristotle.* Oxford University Press, Oxford

Sedley, D. 1991. Is Aristotle's Teleology Anthropocentric? In *Phronesis* 36: 179–96.

2000. Metaphysics, Lambda 10. In Charles and Frede. 2000: 326–50.

2007. *Creationism and Its Critics in Antiquity.* University of California Press, Berkeley and Los Angeles.

2010. Teleology, Aristotelian and Platonic. In Lennox and Bolton 2010: 5–29.

Sharples, R. W. 2004. *Philosophy and the Sciences in Antiquity*, Ashgate, London.

Solmsen, F. 1961. Greek Philosophy and the Discovery of the Nerves. *Museum Helveticum* 18: 150–97.

Sorabji, R. 1980. *Necessity, Cause and Blame: Perspectives on Aristotle's Theory*, Cornell University Press, Ithaca (NY).

Steckerl, F. 1958. *The Fragments of Praxagoras of Cos and His School.* Brill, Leiden.

van der Eijk, Ph. 2005. *Medicine and Philosophy in Classical Antiquity.* Cambridge University Press, Cambridge.

Tricot, J. 1970 *Aristote, Organon 4: Les Seconds Analytiques.* Vrin, Paris.

van Oppenraaij, A. M. I. 1992. *Aristotle De animalibus: Michael Scot's Arabic-Latin Translation. Part 3: Books XV–XIX: Generation of Animals.* Brill, Leiden.

von Staden, H. 1989. *Herophilus: The Art of Medicine in Early Alexandria*, Cambridge University Press, Cambridge.

Ward, J. 1996. *Feminism and Ancient Philosophy.* Routledge, London and New York

Wians, W. 1996. *Philosophical Development, Problems and Prospects.* Lanham, Rowman & Littlefield Publishers, Lanhman (MD) and London.

Witt, C. 2012. Aristotle on Deformed Animal Kinds. In *Oxford Studies in Ancient Philosophy* 43: 83–106.

General Index

Index of Passages